Concurrent, Parallel and Distributed Computing

CONCURRENT, PARALLEL AND DISTRIBUTED COMPUTING

Edited by:

Adele Kuzmiakova

ARCLER
PRESS

www.arclerpress.com

Concurrent, Parallel and Distributed Computing

Adele Kuzmiakova

Arcler Press
224 Shoreacres Road
Burlington, ON L7L 2H2
Canada
www.arclerpress.com
Email: orders@arclereducation.com

ISBN: 978-1-77469-448-0 (Hardcover)

ABOUT THE EDITOR

Adele Kuzmiakova is a machine learning engineer focusing on solving problems in machine learning, deep learning, and computer vision. Adele currently works as a senior machine learning engineer at Ifolor focusing on creating engaging photo stories and products. Adele attended Cornell University in New York, United States for her undergraduate studies. She studied engineering with a focus on applied math. Some of the deep learning problems Adele worked on include predicting air quality from public webcams, developing a real-time human movement tracking, and using 3D computer vision to create 3D avatars from selfies in order to bring online clothes shopping closer to reality. She is also passionate about exchanging ideas and inspiring other people and acted as a workshop organizer at Women in Data Science conference in Geneva, Switzerland.

TABLE OF CONTENTS

LIST OF FIGURES

LIST OF ABBREVIATIONS

3D	three-dimensional
AI	artificial intelligence
AWS	Amazon web services
BBWN	bisection bandwidth per node
CORBA	common object request broker architecture
COTS	commodities off-the-shelf
CPU	central processing unit
CRCW-PRAM	concurrent read concurrent write PRAM
DCE	distributed computing environment
DCOM	distributed component object model
EC2	elastic compute cloud
EREW-PRAM	exclusive read exclusive write PRAM
FIFO	first-in-first-out
FLIT	flow control digit
GPU	graphics processing unit
ISPs	internet service providers
JMS	Java mobile-based social
LAN	local area network
LMM	local memory machine
MIMD	multiple instruction multiple data
MISD	multiple instruction single data
ML	machine learning
MMM	modular memory machine
MOM	message-oriented middleware
MPI	message passing interface
NFS	network file system
NFSM	non-deterministic FSM
NOW	network of workstations
PDA	pushdown automata
PRAM	parallel random-access memory

PVM	parallel virtual machine
RAM	random access machine
RMI	remote method invocation
ROI	remote object invocation
RPC	remote procedure call
SGML	standardized general programming language
SIMD	single instruction multiple data
SISD	single instruction single data
SOA	service-oriented architecture
SSE	streaming SIMD extensions
TAS	test-and-set
TLA	temporal logic of actions
TMR	triple modular redundancy
WAN	wide-area network
WMB	WebSphere messaging broker
WWW	world wide web

PREFACE

Computing that is concurrent, parallel, and distributed (CPDC) is increasingly integrated into nearly all aspects of computer use. It is no longer adequate for programmers of any level to have merely abilities in conventional sequential programming. Even beginner programmers need to be able to do more. This book discusses hands-on exploration of parallelism, targeting instructors of students with no or minimal prior programming experience. The hands-on activities are developed in the Scratch programming language. Scratch is used at various higher-education institutions as a first language for both majors and nonmajors. One advantage of Scratch is that programming is done by dragging and connecting blocks; therefore, students can create interesting programs extremely quickly. This is a crucial feature for the prerequisite-free approach.

It is becoming increasingly important to do computer activities "in parallel," also known as concurrently, since the availability of large amounts of data and the number of people using the internet at the same time continues to expand simultaneously. The concepts of parallel and distributed computing are applicable to a wide variety of subfields within the field of computer science. These subfields include algorithms, computer architecture, networks, operating systems, and software engineering. At the beginning of the 21st century, there was a meteoric rise in the design of multiprocessors as well as other tactics for making complicated programs run more quickly. The ideas of concurrency, mutual exclusion, consistency in state and memory manipulation, message transmission, and shared-memory models constitute the foundation upon which parallel and distributed computing is built.

The book contains multiple small examples, demonstration materials, and sample exercises that can be used by instructors to teach parallel programming concepts to students just being introduced to basic programming concepts. More specifically, it addresses Python multiprocessing features such as fork/join threading, message passing, sharing resources between threads, and using locks. Examples of the utility of parallelism are drawn from application areas such as searching, sorting, simulations, and image processing.

The book is divided up into a total of eight different chapters. The first chapter provides the readers with an in-depth introduction to computing models. The development of concurrent, parallel, and distributed computing is the topic of discussion in Chapter 2. The concurrent computing concept is discussed in great detail in Chapter 3.

The readers are given a comprehensive introduction to the parallel computing concept in Chapter 4. The distributed computing model is the primary topic of discussion in Chapter 5. The applications of concurrent computing, parallel computing, and distributed computing are illustrated in Chapter 6. The most recent advancements in computing models are discussed in Chapter 7. In the last chapter, "Concurrent, Parallel, and Distributed Computing," we discuss the prospects for these three types of computing.

This book does a great job of providing an overview of the many different aspects of the computer field. The material is presented in the book in such a way that even a reader with no prior knowledge of computers may understand it and get familiar with the fundamental ideas involved.

CHAPTER 1

FUNDAMENTALS OF CONCURRENT, PARALLEL, AND DISTRIBUTED COMPUTING

CONTENTS

1.1. INTRODUCTION

The invention of the electronic digital computer is widely regarded as among the most significant advances of the 20th century. Computers and communications technology brought about significant changes in the realms of business, heritage, administration, and scientific knowledge, and has had an impact on virtually every aspect of life (Kshemkalyani and Singhal, 2011). This work provides an overview of the topic of computing and delves into the underlying ideas and procedures that are integral to the creation of software programs that run on computers. Working in a foreign nation for the first time is similar in some ways to begin a career in a whole new sector, such as computer programming. New arrivals to a nation may experience feelings of disorientation and even be rendered helpless due to the significant differences that exist between nations even though all nations share certain fundamental characteristics, such as the requirement for speech and the proclivities for society and start trading (Raynal, 2016). Additionally, it is hard to even identify the characteristics of a particular nation in a manner that is final since such characteristics vary from region to region and change a lot of time. In a like fashion, joining the area of computers may be unsettling, and it can be difficult to find precise descriptions of the aspects of the discipline (Duncan, 1990).

Nevertheless, the discipline of computers is built on a foundation of basic principles that can be stated, taught, and put to use efficiently. Every aspect of computing is predicated on the organized use of networked computers, which are referred to as hardware, as well as the computer algorithms which drive them, which are referred to as software (Fujimoto, 2000). Furthermore, all application programs are constructed utilizing data processing requirements, which are referred to as data structures. Notwithstanding the continuous advancement of software and hardware components as well as the ongoing generation of different frameworks for data processing specifications, these basics have stayed extremely consistent throughout the history of technology. This is the case although newer technologies are being introduced (Raynal, 2015).

In computing curriculum content 2005: The Outline Report, which was created by a working group of ACM, AIS, and IEEE. The following description was provided: "In a broad approach, we may describe computing to include any main objective activity needing, benefitting from, or generating computers. This is a fairly wide concept that includes the usage of computer applications as well as the production of computer software

(Burckhardt, 2014). Additionally, it also includes the creation of computer hardware. This chapter concentrates on the most recent of these business endeavors, which is the creation of computer software. In addition, there are forms of computing that have developed throughout the period to be used in computing. Concurrent programming, parallel processing, and distributed computing are the three paradigms included here (Fujimoto, 2000).

1.2. OVERVIEW OF CONCURRENT COMPUTING

The capability of separate portions or components of a program, method, or issue to be performed in or out of the linear interpolation without impacting the conclusion is referred to as simultaneous computer science or parallelism. This permits contemporaneous units to be executed in parallel, which may considerably enhance total execution speed in inter and multi-core computers. Multi-threading, in more technical words, is the capacity of a program, method, or problem to be decomposed into fact necessary or partly pieces or units (Figure 1.1) (Teich et al., 2011).

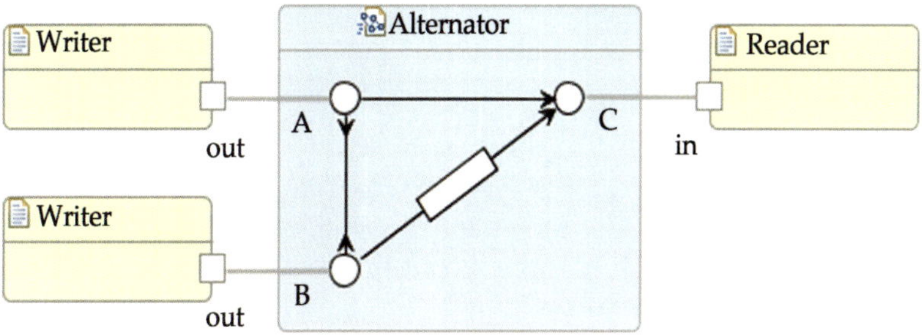

Figure 1.1. Circuito de Rio.

Source: https://computingstudy.wordpress.com/concurrent-parallel-and-distributed-systems/.

Fuzzy systems, processes calculi, the concurrent accidental – acccss memory device model, the actors perfect, as well as the Reo Coordinating Language are some of the statistical equations that have been created for broad concurrent processing. "While concurrent policy coherence had been contemplated for decades, the computer programming of concurrency started with Edsger Dijkstra's famous 1965 article that presented the deadlock issue," writes Leslie Lamport (2015). Within centuries thereafter, there has been a massive increase in interest in concurrently, especially in distributed applications. Thinking back to the field's beginnings, Edsger Dijkstra's pivotal influence shines out" (Jannarone, 1993).

The range of alternative execution routes in education incorporation may be extraordinarily enormous, and the resultant conclusion might be unpredictable since calculations in a concurrent system could engage with one another while being performed. Concurrent usage of resource sharing may cause indecision, resulting in deadlocks with resource scarcity. Developing dependable strategies for combining their operation, data interchange, memory management, and implementation planning to reduce reaction time and maximize capacity are often required while designing concurrent applications (Sunderam et al., 1994).

Theoretically, computer programming has long been interested in the concurrency concept. Carl Adam Petri's pioneering work upon Petri Nets in the 1960s was among the earliest ideas. Within years afterwards, a broad range of conceptual frameworks for describing and understanding concurrently have indeed been created (Chen et al., 2002).

Concurrent measures are developed to the languages and methods that are used to create concurrent applications. Concurrent software is sometimes seen as being broader than parallel processing since it may incorporate random and variable connections and patterns of interaction, while serial systems normally have a specified and very well communication process. Correctness, speed, and resilience are the primary aims of parallel processing. Concurrent programs, such as computer organizations and folder systems, are often intended to run forever, with automated recovery in the event of a malfunction, and not to shut down suddenly. Various concurrent algorithms use transparency parallelism, in which several computing entities fight for something and share the common resource while the developer is unaware of the complexity of the struggle and cooperation (Hansen, 2013).

Concurrent networks generally need the presence of some form of arbitration somewhere within their implementations (typically in the physical infrastructure) to manage access to network resources since they employ common capitals. The employment of arbitrators raises the prospect of uncertainty in simultaneous computing, which has significant ramifications for accuracy and performance in application. Mediation, for example, generates unlimited nondeterminism, which complicates model testing since it produces a phase space expansion, potentially resulting in structures with an unlimited lot of states. Co-processes and predictable concurrency are two concurrent computing methods. In such architectures, controlling thread expressly hand over their different activities and tasks to the systems or the other program (Angus et al., 1990).

Concurrency refers to the simultaneous execution of numerous calculations. Whether we like it or not, parallelism is omnipresent in contemporary program development (Aschermann et al., 2018), as indicated by the presence of:

- A system is a group of linked computers;
- Apps which are operating simultaneously on the same machine;
- A system with large multiple processors.

Concurrency is, in fact, critical in contemporary coding (Pomello, 1985), as demonstrated by these facts:

- Sites must be able to manage several users at the same time;
- Several of the programming for mobile applications must be done on databases ("inside the clouds").

A software tool often requires a backstage operation that does not disrupt the user. Horizon, for example, constructs your Programming language as you operate on it (Agha and Hewitt, 1987).

Parallel processing will remain important in the future. Maximum clock rates are not increasing anymore. Rather, we get more processing with each new generation of contemporary CPUs. As a result, shortly, we will have to partition a computation into many concurrent bits to keep it running faster (Fernandez et al., 2012).

1.2.1. Two Models for Concurrent Programming

Simultaneous processing is usually done using one of two approaches: identify a specific or message forwarding (Figure 1.2) (Ranganath and Hatcliff, 2006).

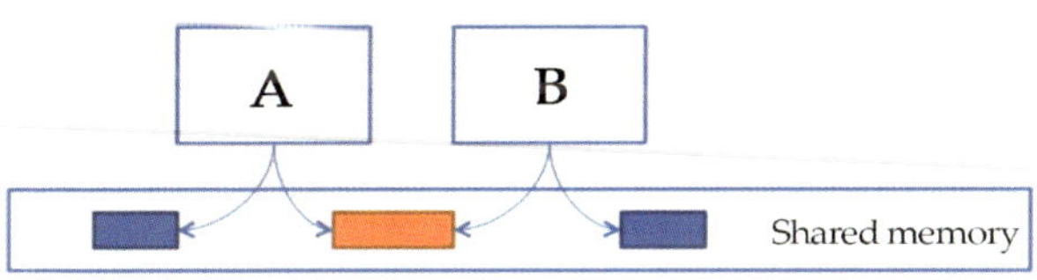

Figure 1.2. Memories that are linked.

Source: https://web.mit.edu/6.005/www/fa14/classes/17-concurrency/#:~:text=There%20are%20two%20common%20models,shared%20memory%20and%20message%20passing.

- **Shared Memory:** Simultaneous modules communicate in the sharing memory system concurrently by learning to write sharing variables in storage (Kasten and Herrmann, 2019).

Other applications of the ability to share concepts include:

- A and B could be two processing (or processing cores) wanting to share memory space in the very same machine.
- A and B could be two applications that operate on the same machine and share a shared filesystem containing files that they can deliver and understand.
- A and B could be two filaments in a very similar Java application, exchanging the same Java objects (we will describe this process is later) (Figure 1.3).

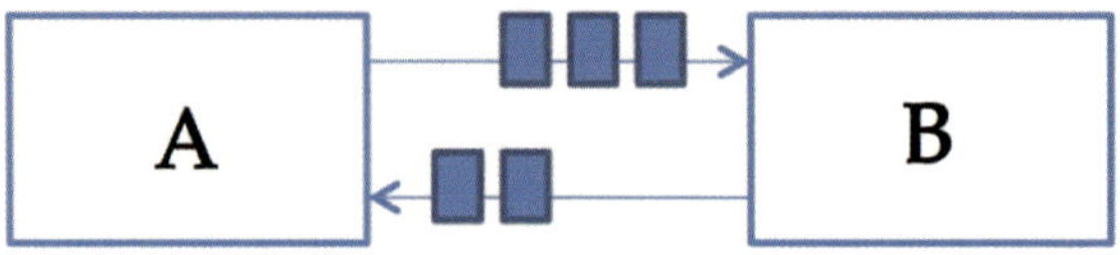

Figure 1.3. Text transmission.

Source: https://web.mit.edu/6.005/www/fa14/classes/17-concurrency/#:~:text=There%20are%20two%20common%20models,shared%20memory%20and%20message%20passing.

- **Message Passing:** Parallel modules communicate in the text paradigm by sending texts to each other across a channel of communication. Communications are sent from modules and scheduled for processing. Here are several examples (Harris et al., 2007):
  - Both A and B could be two workstations connected to a network through internet connections.
  - Both A and B may be a browser extension and just a net host, in which case A connects to B, requests a website, and B returns the data from the web page to A.
  - A and B could be a customer and server for an instant messenger.
  - Both AB could be two programs that run along with the same computer with a pipe connecting their output and input, such as ls | grep entered into the command line.

1.2.2. Processes, Threads, Time-Slicing

Parallel programs interact using the text and sharable paradigms. Parallel modules are divided into two categories: processors and threads. Process A is a separate version of a computer system on the same system that is segregated from other processes. The process has its own isolated space in the computer's retention (Balling and Sobieszczanski-Sobieski, 1996).

A virtual computer represents the processing abstraction. It gives the impression that the software has the whole system to itself as if a new computer with a concise overview was built only to execute that program. Programs, like computers linked across a connection, typically do not share memory. A process cannot access the storage or objects of another process. On most operating systems, pooling memory across processes is feasible, but it requires extra work. A new protocol, on the other hand, is prepared for data transmission right away since it is built with standardized incoming and outgoing streams, like the integrated self and system (Park et al., 2006). These streams use Java as the main programming language (Fox, 1988).

Although threads share all of the information inside the system, they are inherently prepared for shared memory. Obtaining "thread-local" memory that is exclusive to a separate thread takes extra work. The text must also be expressly set up by establishing and utilizing queue database systems. In a later reading, we will take a deep dive into how to achieve it (Figure 1.4) (Gulisano et al., 2015).

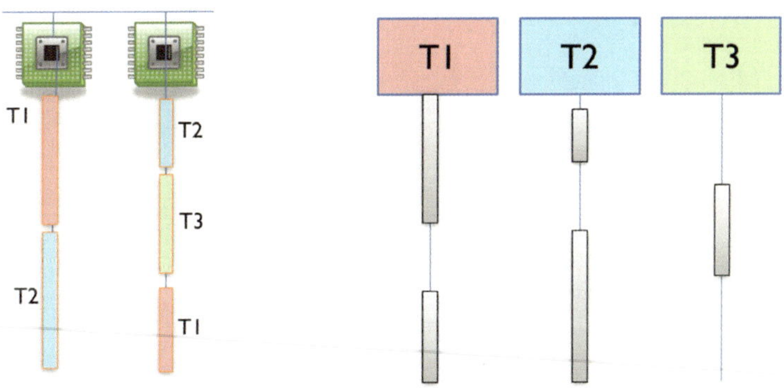

Figure 1.4. Time slicing.

Source: https://docs.oracle.com/javase/tutorial/essential/concurrency/procthread.html.

With just between one-two processing unit in my machine, how can I run several concurrent strands? Concurrency is emulated through time cutting, which implies the system alternates between threads because there is more threading than the processor. On a computer with just processor cores, three threading T1, T2, and T3 may be time-sliced as shown in the diagram on the right (Geist and Sunderam, 1992). Time moves downhill in the diagram, so one CPU runs strand T1 while the extra runs strand T2, and afterwards the second machine changes to strand T3. Strand T2 merely stops until its next time slice, which might be on the same CPU or a different processor.

For most devices, instant cutting is unexpected and planned, which implies a thread can be halted or resumed at any time (Walker, 1994).

1.2.3. Shared Memory Example

Let us have a look at a distributed recollection system in action. The purpose of such instance is to demonstrate that simultaneous programming is difficult due to the possibility of subtle flaws (Figure 1.5) (Futatsugi and Nakagawa, 1997).

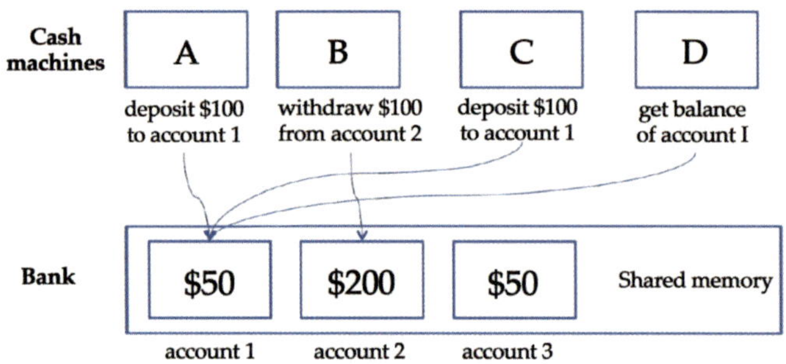

Figure 1.5. Example of a shared memory.

Source: https://web.mit.edu/6.005/www/fa14/classes/20-queues-locks/synchronization/.

Consider a set with cash registers that employ a recollection paradigm, which allows all of the devices to read English the same account items in storage (Stoughton and Miekle, 1988).

// Assume that all of the cash points have the same checking account.

private static int balance = 0;

```
private static void deposit() {
balance = balance + 1;
}
private static void withdraw() {
balance = balance – 1;
}
```

Clients just use cash machines that make the company must follow (Ishikawa and Tokoro, 1986):

```
deposit(); // put a dollar in
withdraw(); // take it back out
// each ATM does a bunch of transactions that
// modify balance, but leave it unchanged afterward
private static void cashMachine() {
for (int i = 0; i < TRANSACTIONS_PER_MACHINE; ++i) {
deposit(); // put a dollar in
withdraw(); // take it back out
}
}
```

Irrespective of how many cash points were operating or even how many trades we handled, the checking account should remain nothing after the day.

However, when we execute this code, we commonly realize that the ending balance of the day is not zero. If many than a cash machine() function is executing at the same time – for example, on different processors in the very same machine – the ending balance of the day may not have been zero. What is to stop you (McBryan, 1994)?

1.2.4. Concurrency Is Hard to Test and Debug

However, the concurrency is inherently difficult because it is not very straightforward to identify racing circumstances via experimentation. Even if a test finds a flaw, pinpointing the section of the software that is producing it might be difficult (Chen et al., 2021).

Concurrency difficulties are notoriously hard to replicate. It is tough to get them to occur in the same method recurrently. The comparative accuracy of predictions that are highly impacted by the atmosphere determines how commands or information are interleaved. Additional running applications, other network activity, system software development selections, differences in computer timepiece speed, and so on may all generate stays. You could obtain different results each time you execute a program with compatibility issues (Benner and Montry, 1986).

These are heisenbugs, which are difficult to replicate, as contrasted to "bohrbugs," which appear every time you look at them. Bohrbugs account for almost all bugs in linear development.

Concurrency is difficult to master. However, do not let this deter you. We will examine rational techniques to build concurrent software that is secure from such types of issues during the following many readings (Fusco, 1970).

1.3. OVERVIEW OF PARALLEL COMPUTING

Functionality is a kind of computing that involves running many calculations or activities at the same time. Large problems are frequently broken down into smaller problems that can be addressed at the same time. Bit-level, direction, information, and activities parallelism are the four forms of parallel processing. Parallel processing has been around for a while, especially in high-end computing, but it has lately gained traction because of physical limits that constrain speed development. Because energy usage (and therefore heat generation) by computing has become a concern in recent years, data processing has been the dominant paradigm in computer coding, particularly in the case of inter (Park, 2008).

Simultaneous computing and simultaneous computing are commonly confused and used interchangeably, although the two are distinguishable: parallelism may exist without concurrency (including smidge concurrency), and concurrently can exist with parallel processing (like multitasking by time-sharing on a single-core CPU) (Figure 1.6) (Thanisch, 2000).

Figure 1.6. IBM's massively parallel supercomputer blue gene/P.

Source: https://computingstudy.wordpress.com/concurrent-parallel-and-distributed-systems/.

Processer effort is frequently subdivided into numerous, sometimes many, very like subprojects that may be performed separately but whose outputs are then joint after the conclusion in similar processing. In comparison, in concurrent computers, the numerous processes seldom handle linked tasks; if they do, as in cloud applications, the individual tasks are generally diverse and need some inter-process interaction during the performance (Alistarh, 2020). Multi-core or multi-processor systems have numerous dispensation parts inside a single system, while clustered, MPPs, and grids operate on the very same job using several processers. For speeding certain activities, customized parallel processing architecture is often employed alongside regular CPUs. In certain cases, including even bit-level or guidance parallel processing, parallelism is translucent to the developer, but expressly parallel methodologies, especially those using parallelization, are harder to write than sequence numbers, even though concurrency tries to introduce a few new categories of possible future software bugs, a most popular of whom are race conditions. Communications and synchronization between sub-tasks are often the most difficult aspects of achieving effective concurrent program effectiveness (Fournet et al., 2000).

Splitting down large processes into tiny, autonomously, and often connected pieces that can be executed simultaneously by numerous processors communicating via memory locations, with the outputs being collected as part of the full method, is known as system computing. The basic goal of parallel processing is to increase available processing power for faster submission of applications and problem resolution (Yonezawa, 1994).

The web application distributes data processing applications in smaller pieces, which are then placed through a process upon every server. Perpendicularly computing facilities are generally housed inside a single data center, in which many processing is placed in a computer system; data processing requests are dispersed in smaller pieces by the database server, which are then placed through a process on every server.

Bit-level parallel, guidance parallelism, project correspondence, and wonderful word-level parallel processing are the four procedures of parallel computation accessible from both isolated and open access massively parallel vendors (Vajteršic et al., 2009).

Bit-level concurrency improves computer word size, reducing the number of commands processing needed to act on variables larger than the continuous speech.

Parallel processing at the beginning of a lesson: the technology method uses dynamically parallelism, wherein the processor determines which commands to perform in simultaneous at real-time; the programming approach uses statically parallelism, wherein the compiler selects which commands to perform in parallel.

Task parallelism: a way of running multiple computer codes over several processors to perform real-time actions along with the same data (Paprzycki and Zalewski, 1997).

Super word-level parallelism: a segmentation method that makes use of parallel coding correspondence. Powdered concurrency, wherein subtasks interact many areas per second; poorly graded parallel processing, wherein sub-processes do not interact multiple times a second; or chastening parallel processing, wherein subtasks rarely or never interrelate, are the three types of parallel programs. Mapping is a technique for solving excruciatingly equivalent issues in similar processing by performing a single process on all items of a series without needing interaction among sub-tasks (Ostrouchov, 1987).

Equivalent computation became general and changed in the 21st period as a result of processing performance grading reaching its boundary. To solve the problem of energy consumption and overheated CPU cores, developers, and producers started building corresponding system software programs and manufacturing high well-organized processors with several cores (Yang et al., 2021).

Through the amplified use of multiple cores central processing units (CPU) and GPUs, equivalent computing is flattering more important. CPUs and GPUs work composed to improve information speed and the number of concurrent calculations confidential a request. A GPU can finish more tasks in an assumed period than just a CPU thanks to concurrency (Figure 1.7) (Barney, 2010).

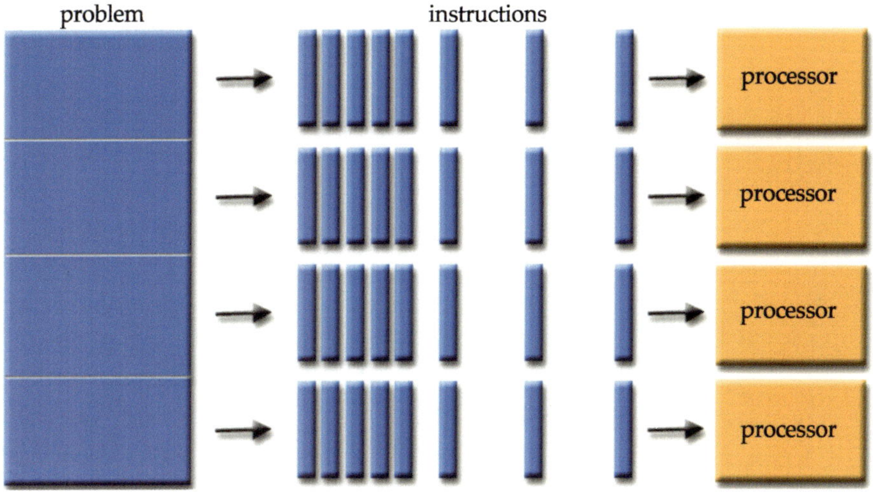

Figure 1.7. A typical example of parallel processing.

Source: https://hpc.llnl.gov/documentation/tutorials/introduction-parallel-computing-tutorial.

1.3.1. Fundamentals of Parallel Computer Architecture

Parallelization design can be found in various parallel processing, which is classified based on the number of parallel processors they can handle. Concurrent computer networks and coding methodologies work together just to maximize the capabilities of such devices. Parallelization systems may be seen in illustrations (Saidu et al., 2015):

- **Multi-Core Computing:** A inter microprocessor is an electrical component of a computing device that has two or maybe more core processors that perform instructions sets in parallel.

Action, vectors, and parallel processing architectures are multi-threading units which are executed on many dies within a chip packing or a pure silicon device die. Multi-core systems may be heterogeneous, with homogeneity design containing only cores that are identical and inhomogeneous designs include cores that are not comparable.

- **Symmetric Multiprocessing:** In a parallel processing software/hardware architecture, two or much more unbiased, largely homogeneous computations are run by a single operating system instance it wants to serve all computing power equitably and is connected to a solitary, conveyed memory area with total access to all sharing of resources and devices. Every CPU will have its system memory, can interface with some other processing through it on data networks, and can perform any task irrespective of where the necessary data is stored in storage.
- **Distributed Computing:** The program's elements are dispersed over many computer systems, communicating using simple HTTP, RPC-like links, and important moves to coordinate their functions. Distributed systems have two crucial characteristics: machine breakdowns tolerance and component concurrent. The most prevalent parallelization designs include client, three-tier, n-tier, and interpersonal designs. There is a lot of overlap in parallelization systems, and the expressions are commonly used interchangeably.
- **Massively Parallel Computing:** It relates to the use of large multiple processors or mainframe computers to do a sequence of computations at the very same time. One option is to group several computers into a centralized, officially structured computer system. Another method is a public cloud, wherein numerous widely dispersed computers cooperate and communicate over the Internet to solve a given problem.

Dedicated parallel computers, clump computing, distributed systems, velocity processors, implementation circuits, general-purpose information technology on the graphics processing unit (GPU (GPGPU), and reprogrammable computing to practice area components are some of the other parallel computer frameworks. Spread memory or memory location is the type of memory in just about any simultaneous computer system (Almási, 1985).

1.3.2. Parallel Computing Software Solutions and Techniques

To make parallel processing on simultaneous hardware easier, concurrent application programs, libraries, APIs, and parallel processing paradigms have been created. The following are some examples of parallel processing software and methods:

- **Application Checkpointing:** A system software issue detection technique that stores every one of the company's various states and enables the program to continue and restart with such a point in the case of a problem. Multithreaded control is a critical component of parallel processing software programs that spread operations over a huge number of processors (Tasora et al., 2011).
- **Automatic Parallelization:** It relates to the translation of sequential code into several co-codes to fully use much processing at the very same time in an allowed multiprocessors (SMP) system. Automatic parallelization techniques include parsing, analyzing data, planning, and optimizations. The Pyramid translator, Polaris compiled code, Paddy Fortran D optimization, SUIF computer program, and Vienna Fortran optimization method are all examples of guidelines that enable a compiled code and technologies more effective (Ohbuchi, 1985).
- **Parallel Programming Languages:** Person takes and distributed memory programming are the second most frequent kinds of parallelization languages. Sharing recollection scripting languages communicate by modifying memory settings, while shared storage method computing needs a communicated signal.

1.3.3. Difference Between Parallel Computing and Cloud Computing

Cloud technology is a broad word that refers to the on-demand, pay-as-you-go provision of modular programs and reusable code, including databases, storage systems, connectivity, hosts, and software through the Web (Hutchinson et al., 2002).

Cloud services may be publicly or privately, are completely controlled by the supplier, and allow for distant access to the information, activities, and apps from every internet-enabled device. The three most common service kinds are Infrastructure as a Service (IaaS), Cloud solution (PaaS), and Cloud computing (SaaS).

Technology is a relatively new revolution in an operating system designed to allow for the availability of wide parallel processing via vast, virtual machine clusters, having allowed the typical consumer and relatively small institutions to take advantage of parallelization processing and memory options previously reserved for large companies (Ligon et al., 1999).

1.3.4. Difference Between Parallel Processing and Parallel Computing

Concurrent computing is a method of computation that divides and processes discrete pieces of a more complex activity on multiple CPUs at the very same effort, saving time (Valentini et al., 2013).

Utilizing parallelization software programs, software developers split and assign each task to a distinct processor, which should operate to compile and read the stuff when each processor has finished its calculation. This operation is carried out using a computer network or a system with two or more processor units. Computer-based systems and comparison handling are often used interchangeably because they occur simultaneously; however, whereas parallel processing shows the number of cores or CPUs running concurrently in a framework, multitasking relates to the manner software works to optimize for that situation (Treleaven, 1988).

1.3.5. Difference Between Sequential and Parallel Computing

Sequencing computation, often referred to as sequential computation, represents a single processor to run a computer program. The program is divided into a series of individual commands that are performed in rapid succession with really no interruption. The software has typically been developed in a systematic fashion, which is a more straightforward technique, but is constrained by the system's speed and capacity to perform each sequence of instructions. Whereas sequential patterns are used on single-processor devices, concurrent data patterns are used in parallel processing settings (Fryza et al., 2012).

Performance measurement in sequentially coding is significantly less complicated and crucial than benchmarking in parallel computing since it usually simply entails detecting system constraints. Benchmark and performing testing platforms, which leverage a range of measuring approaches, including statistical techniques and many repetitions, may be used to obtain benchmarks in parallel computing. Parallel computing for data scientists, machines learning distributed processing, and parallel computing artificially intelligent use applications all benefit from the opportunity to sidestep this bottleneck by transferring data around the system memory (Sérot and Ginhac, 2002).

Parallel computing is the polar opposite of serial computing. Although parallel computing is more difficult and expensive up front, the benefit of just being able to solve the problem quicker often surpasses the expense of serial computing technology (Parashar and Hariri, 2004).

1.4. OVERVIEW DISTRIBUTED COMPUTING

Cloud computing is a part of computer science concerned with distributed systems. A database system is a method for information systems components to interact and control their operations via the use of messages. To reach a shared purpose, the components communicate with one another. Part parallelism, the absence of a universal clock, and individual hardware failures are three essential features of distributed systems. Service level agreement networks, hugely online play games, and mentoring systems are all examples of data warehousing (Teich et al., 2011). Cloud computing is an area of computer science concerned with distributed systems. A database management system is a method for central computer elements to interact and manage their operations via the use of messages. To reach a shared purpose, the components interact with one another. Part parallelism, the absence of a universal clock, or individual hardware failures are three essential features of distributed applications (Qian et al., 2009). Software as a service networking, hugely multiplayer play games, and mentoring systems are all examples of data warehousing. Decentralized technologies are computing model that works on a distributed network, and dispersion programming is the process to develop them. To mention a few, message delivery mechanisms include genuine HTTP, RPC-like adaptors, and interaction (Singh, 2019).

The following are some of the reasons why shared networks and cloud computing are used:

- Features of an inquiry Genus Laevigata necessitates the installation of a telecommunications network that connects several processors, including figures made inside one physical address but required in another.
- In several circumstances, using a computer system might be feasible in theory, but the usage of a distributed network is advantageous for pragmatic purposes. In comparison to

a particular high processor, it will be more cost-effective to achieve the necessary performance level by employing a cluster of numerous low-end systems. Because there is no weak point in a distributed network, it may be more reliable than a non-distributed framework. Furthermore, compared to a monolithic single processor system, a distributed database system could be simpler to grow and administer.

It should be the procedure of linking many processer networks to form a group via a system to transfer data and coordinate computing power. A "distributed database system" is the name given to such a group. Flexibility (through a "scalability design"), efficiency (via parallel), robustness (by redundancies), and expense are all benefits of cloud applications (through the use of low-cost, commodity hardware) (Kendall et al., 2000).

As data volumes have grown and application performance criteria have increased, cloud computing has become more common in database and application design. That was why flexibility is so crucial since as the amount of data grows, the extra strain can be readily handled by adding more hardware to the connection. Traditional "big iron" configurations, which rely mostly on strong CPUs, must cope with load spikes by updating and replacing gear (Figure 1.8) (Oduro et al., 1997).

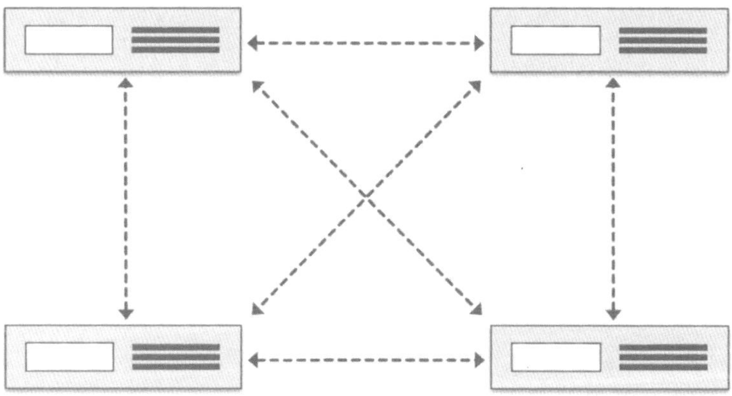

Figure 1.8. Multi-user computer systems are linked together via a network in cloud applications to allow enormous tasks to make use of all resources available.

Source: https://hazelcast.com/glossary/distributed-computing/.

1.4.1. Distributed Computing in Cloud Computing

Due to the growing number of cloud providers, cloud computing technology became even more accessible. Although cloud storage installations do not support cloud apps by default, there seems to be a variety of dispersed software packages that may operate on the side to make use of the easily available processing capacity (Zhang et al., 2020).

Companies traditionally relied on data analysts (DBAs) or technology suppliers to link computer resources across networks within and outside data centers to get data. Many cloud providers are already making it simple to add computers to a group to expand capacity or processing capacity (Shrivastava et al., 1989).

Cloud technology allows for better degrees of flexibility when dealing with expanding workloads due to the simplicity and speed in which additional computing resources may be deployed. This allows for "elasticity," where a group of systems may readily grow or shrink in response to changing workload demands (Martins et al., 2012).

1.4.2. Key Advantages of Distributed Computing

- **Cloud Technology:** It enables a set of systems to work together like they were the same. This provides several benefits.
- **Scalability:** Scaling distributed software groups is simple because of a "scalability design," which allows for bigger loads to be managed by adding extra hardware (versus replacing existing hardware).
- **Performance:** Through a start dividing technique, the clusters may reach high performance by using parallelism, where each processor in the group tackles a portion of an aggregate job at the same time.
- **Resilience:** Concurrent programming groupings often replicate or "duplicate entries" data among all computing systems to ensure that there is no weak point. When a computer breaks, backups of its information are made and stored elsewhere to guarantee that neither error occur.

- **Cost-Effectiveness:** Cloud computing frequently makes use of preexisting hardware, reducing the cost from both original installs and subsequent expansion.

A dispersed network is a combination of separate systems and global data storage components. Such elements may interact, coordinate, and work together to complete the same goal, creating the appearance of a single, universal system with a strong processing capacity (Wu et al., 2016).

Distributed systems include distributed computer servers, databases, application software, and file storage solutions (Pursula, 1999).

1.4.3. Examples of Distributed Systems

- The world wide web (WWW) itself;
- A communication system with many antennas, amplification, and other network components to create a unified system.

1.4.4. How Does Distributed Computing Work?

Distributed computing brings together software and hardware to accomplish a variety of tasks, which include (Kan et al., 2019):

- Working together to accomplish a common objective via resource sharing;
- Handling multiple access privileges based on their authority level;
- Maintaining open-source components, such as distributed system software, for future growth;
- Working on the same task simultaneously;
- Verifying that all computer resources are accessible and that several machines may run at the same time;
- Identifying and managing faults in the distribution network's linked parts such that the connection does not fail.

Automated procedures and APIs are used in plugin adds systems to improve performance (Jimack and Topping, 1999).

Distributed clouds are beneficial for organizations in terms of flexibility. On-premises technologies may be connected to the cloud services stack by cloud providers, allowing businesses to modernize their whole IT network without having to destroy outdated arrangements. Conversely, they may use infrastructure to enhance it with minimal changes (Kirk, 2007).

The cloud provider is in charge of dispersed infrastructure's application updates, security, dependability, standards, administration, and disaster response method (Motta et al., 2012).

1.4.5. What Are the Advantages of Distributed Cloud Computing?

Distributed compute clusters have become a fundamental offering which all cloud computing providers offer to their customers. Their benefits are highlighted in subsections.

1.4.5.1. Ultimate Scalability

The distributed nodes or parts are self-contained processors. They constitute a distributed computer cluster when they work together. You may quickly add and remove devices from the connection without putting a load on the channel's resources or causing downtime. It is simple to scale using distributed software nctwork operators (Ierotheou et al., 1996).

1.4.5.2. Improved Fault Tolerance

The distributed system works well together to produce a cohesive system. The design also enables any component to enter or quit at any moment. Failure distributed systems, as a consequence, have a greater level of dependability (Drozdowski, 1996).

1.4.5.3. Boosted Performance and Agility

Distributed clouds enable numerous computers to work on the very same task at the same time, boosting the performance of the system by two orders of magnitude or more. Using distributed applications, sensing performance and expense may increase as a consequence of congestion control (Ahmad et al., 2002).

1.4.5.4. Lower Latency

Businesses might use cloud-based systems destinations to speed up application processing since resources are available internationally. Border computing's reduced latency is combined with the ease of a uniform cloud platform to benefit businesses (Elmsheuser and Di Girolamo, 2019).

1.4.5.5. Helpful in Compliance Implementation

Whether it is for industry or local compatibility, the cloud-based architecture enables enterprises to access local or great-nation services across several regions. This allows them to simply comply with a variety of data privacy rules, with the GDPR in Europe or even the CCPA in California (Howard, 1988).

You could read our post on the benefits of cloud applications if you want to understand more about the advantages of cloud applications (Ali and Khan, 2015).

1.4.6. Four Types of Distributed Systems

There seem to be a few distinct designs that fall under the heading of distributed applications. Cloud-based systems may be divided into four categories in general (Elzeki et al., 2012):

- **Client-Server Model:** The service consumer retrieves data from a server under this paradigm, then prepares and presents it to the client. End-users may immediately upload their updates to the system to amend this data.
  - *For example*, customers' data is stored by firms such as Amazon. When one consumer changes their telephone number, they communicate the information to the server, which upgrades the databases (Figure 1.9) (Divakarla and Kumari, 2010).

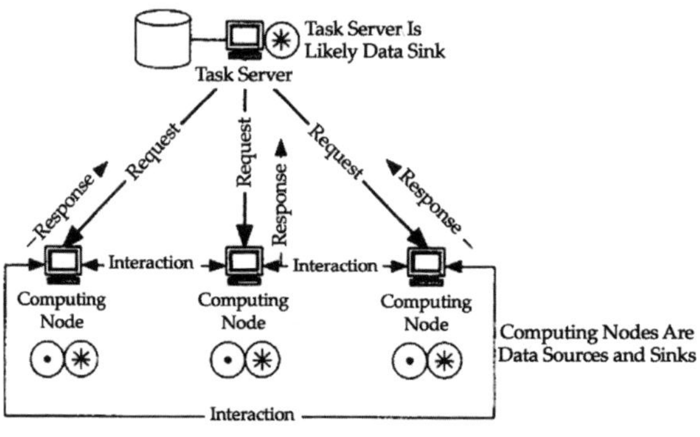

Figure 1.9. A flow model for distributed computing.

Source: https://www.sciencedirect.com/topics/computer-science/distributed-computing.

- **Three-Tier Model:** The agency tier is added to the multiple paradigms, which sits between the server and the client.

The customer data is stored in this intermediate layer, which relieves the customer of the responsibility of data management. Typically, a web application allows the customer to view their data. This reduces and automates the work of the application server as well as the user (Franklin et al., 1992).

- *For example*, a documents editor as well as a cloud storage area store your documents and files. This kind of storage medium may make your file accessible from everywhere over the web, sparing you the period and exertion of maintaining information on your workstation (Singh, 2019).
  - **Multi-Tier Model:** To communicate with numerous backup data levels and front display tiers, businesses need business rules. This concept makes it simple to submit queries to numerous corporate network applications. Large companies favor the n-tier or multi-tier distributed software approach for this reason.
- *For example*, whenever a user uploads a post on social media to numerous platforms, a business network comprising n-tiers collaborates. The post moves from either the data toward the application level (Berman et al., 2003).
  - **Peer-to-Peer Model:** This approach, unlike the conventional client-server model, consists of peers. According to the request being processed, each peer might operate as a consumer or a service. Such peers pool their computational power, judicial authority, and ability to collaborate more effectively (Zaslavsky and Tari, 1998).
- *For example*, Nodes on the chain collaborate to make choices about adding, removing, and modifying data from the network (Shrivastava et al., 1988).

1.4.7. Applications of Distributed Computing

CDNs spread their assets around the world, allowing users to acquire the most recent copy to speed up their searches. The broadcasting and video surveillance businesses benefit much more from certain setups. Whenever a customer in Seattle clicks on the link to the video, the dispersion system routes the request to a local CDN in Washington, allowing the customer to watch the movie faster (Figure 1.10) (Want et al., 1995).

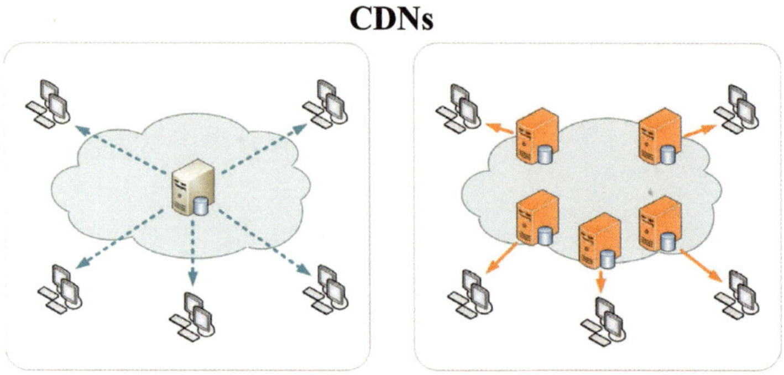

Figure 1.10. Single server distribution (left) and CDN distribution (right).

Source: https://en.wikipedia.org/wiki/Content_delivery_network#/media/File:NCDN_-_CDN.png.

1.4.7.1. Real-Time or Performance-Driven Systems

Distributed machines substantially assist practical applications (those that handle data in a time-critical way), since they must perform quicker via effective data fetching. Actual services where spreading cloud might enhance user experience include games online with significant visual data (e.g., PUBG, and Fortnite), programs with payment choices, and torrenting apps (Parameswaran and Whinston, 2007).

1.4.7.2. Distributed Systems and Cloud Computing

Cloud technology and decentralized systems are a great combination for making networks more efficient and fail. Distributed online services cover all aspects of your organization, from storage to administration (Buyya and Venugopal, 2004).

1.4.7.3. Distributed Computing with Ridge

Organizations may create their unique customized distributed applications using Ridge's dispersed cloud infrastructure, which combines the flexibility of edge computing with the strength of cloud applications (Seymour et al., 2002).

For sophisticated installations, Ridge provides controlled Kubernetes clusters, container management, and data storage resources. Ridge Cloud makes use of localized and transmission economies (Parashar and Lee, 2005).

Ridge clouds, as an alternative to the standard public clouds, basically allow owners to tap into a worldwide network of service providers rather than depending on the availability of computing resources in a single area. It also enables organizations to deploy and endlessly scale apps everywhere they need them by allowing integration with existing infrastructures (Dongarra and Lastovetsky, 2006).

REFERENCES

1. Agha, G., & Hewitt, C., (1987). Actors: A conceptual foundation for concurrent object-oriented programming. In: *Research Directions in Object-Oriented Programming* (Vol. 1, pp. 49–74).
2. Ahmad, I., He, Y., & Liou, M. L., (2002). Video compression with parallel processing. *Parallel Computing, 28*(7, 8), 1039–1078.
3. Ali, M. F., & Khan, R. Z., (2015). Distributed computing: An overview. *International Journal of Advanced Networking and Applications, 7*(1), 2630.
4. Alistarh, D., (2020). Distributed computing column 78: 60 years of mastering concurrent computing through sequential thinking. *ACM SIGACT News, 51*(2), 58.
5. Almási, G. S., (1985). Overview of parallel processing. *Parallel Computing, 2*(3), 191–203.
6. Angus, I. G., Fox, G. C., Kim, J. S., & Walker, D. W., (1990). *Solving Problems on Concurrent Processors* (Vol. 1, 2, pp. 2–5). Prentice-Hall, Inc.
7. Aschermann, M., Dennisen, S., Kraus, P., & Müller, J. P., (2018). LightJason, a highly scalable and concurrent agent framework: Overview and application. In: *AAMAS* (Vol. 1, pp. 1794–1796).
8. Balling, R. J., & Sobieszczanski-Sobieski, J., (1996). Optimization of coupled systems-a critical overview of approaches. *AIAA Journal, 34*(1), 6–17.
9. Barney, B., (2010). *Introduction to Parallel Computing* (Vol. 6, No. 13, p. 10). Lawrence Livermore National Laboratory.
10. Benner, R. E., & Montry, G. R., (1986). *Overview of Preconditioned Conjugate Gradient (PCG) Methods in Concurrent Finite Element Analysis (No. SAND-85-2727)* (Vol. 1, pp. 2–5). Sandia National Labs., Albuquerque, NM (USA).
11. Berman, F., Fox, G., & Hey, T. J. M. G. I. R., (2003). Overview of the book: Grid computing–making the global infrastructure a reality. *Making the Global Infrastructure a Reality, 3*, 2–8.
12. Burckhardt, S., (2014). Principles of eventual consistency. *Foundations and Trends® in Programming Languages, 1*(1, 2), 1–150.
13. Buyya, R., & Venugopal, S., (2004). The Gridbus toolkit for service-oriented grid and utility computing: An overview and status report.

In: *1st IEEE International Workshop on Grid Economics and Business Models, 2004: GECON 2004*. (Vol. 1, pp. 19–66). IEEE.

14. Chen, K., Zhang, D., Yao, L., Guo, B., Yu, Z., & Liu, Y., (2021). Deep learning for sensor-based human activity recognition: Overview, challenges, and opportunities. *ACM Computing Surveys (CSUR), 54*(4), 1–40.
15. Chen, Z., Xu, B., & Zhao, J., (2002). An overview of methods for dependence analysis of concurrent programs. *ACM SIGPLAN Notices, 37*(8), 45–52.
16. Divakarla, U., & Kumari, G., (2010). An overview of cloud computing in distributed systems. In: *AIP Conference Proceedings* (Vol. 1324, No. 1, pp. 184–186). American Institute of Physics.
17. Dongarra, J., & Lastovetsky, A., (2006). An overview of heterogeneous high performance and grid computing. *Engineering the Grid: Status and Perspective, 1*, 1–25.
18. Drozdowski, M., (1996). Scheduling multiprocessor tasks—An overview. *European Journal of Operational Research, 94*(2), 215–230.
19. Duncan, R., (1990). A survey of parallel computer architectures. *Computer, 23*(2), 5–16.
20. Elmsheuser, J., & Di Girolamo, A., (2019). Overview of the ATLAS distributed computing system. In: *EPJ Web of Conferences* (Vol. 214, p. 03010). EDP Sciences.
21. Elzeki, O. M., Rashad, M. Z., & Elsoud, M. A., (2012). Overview of scheduling tasks in distributed computing systems. *International Journal of Soft Computing and Engineering, 2*(3), 470–475.
22. Fernandez, A., Peralta, D., Herrera, F., & Benítez, J. M., (2012). An overview of e-learning in cloud computing. In: *Workshop on Learning Technology for Education in Cloud (LTEC'12)* (Vol. 1, pp. 35–46). Springer, Berlin, Heidelberg.
23. Fournet, C., Fessant, F. L., Maranget, L., & Schmitt, A., (2002). JoCaml: A language for concurrent distributed and mobile programming. In: *International School on Advanced Functional Programming* (Vol. 1, pp. 129–158). Springer, Berlin, Heidelberg.
24. Fox, G. C., (1988). The hypercube and the Caltech concurrent computation program: A microcosm of parallel computing. In: *Special Purpose Computers* (Vol. 1, pp. 1–39). Academic Press.

25. Franklin, M., Galil, Z., & Yung, M., (1992). An overview of secure distributed computing. In: *Proceedings of Sequences II, Methods in Communications, Security and Computer Science*.
26. Fryza, T., Svobodova, J., Adamec, F., Marsalek, R., & Prokopec, J., (2012). Overview of parallel platforms for common high-performance computing. *Radioengineering, 21*(1), 436–444.
27. Fujimoto, R. M., (2000). *Parallel and Distributed Simulation Systems* (Vol. 300, pp. 2–9). New York: Wiley.
28. Fusco, V. F., (1970). Invited paper an overview of the high frequency circuit modelling using concurrent processing techniques. *WIT Transactions on Engineering Sciences, 3*, 3–9.
29. Futatsugi, K., & Nakagawa, A., (1997). An overview of CAFE specification environment-an algebraic approach for creating, verifying, and maintaining formal specifications over networks. In: *First IEEE International Conference on Formal Engineering Methods* (Vol. 1, pp. 170–181). IEEE.
30. Geist, G. A., & Sunderam, V. S., (1992). Network-based concurrent computing on the PVM system. *Concurrency: Practice and Experience, 4*(4), 293–311.
31. Gulisano, V., Nikolakopoulos, Y., Papatriantafilou, M., & Tsigas, P., (2015). Data-streaming and concurrent data-object co-design: Overview and algorithmic challenges. *Algorithms, Probability, Networks, and Games, 1*, 242–260.
32. Hansen, P. B., (2013). *The Origin of Concurrent Programming: From Semaphores to Remote Procedure Calls* (Vol. 1, pp. 2–7). Springer Science & Business Media.
33. Harris, T., Cristal, A., Unsal, O. S., Ayguade, E., Gagliardi, F., Smith, B., & Valero, M., (2007). Transactional memory: An overview. *IEEE Micro, 27*(3), 8–29.
34. Howard, J. H., (1988). *An Overview of the Andrew File System* (Vol. 17, pp. 2–9). Carnegie Mellon University, Information Technology Center.
35. Hutchinson, S., Keiter, E., Hoekstra, R., Watts, H., Waters, A., Russo, T., & Bogdan, C., (2002). The Xyce™ parallel electronic simulator: An overview. *Parallel Computing: Advances and Current Issues, 1*, 165–172.

36. Ierotheou, C. S., Johnson, S. P., Cross, M., & Leggett, P. F., (1996). Computer aided parallelization tools (CAPTools)—Conceptual overview and performance on the parallelization of structured mesh codes. *Parallel Computing, 22*(2), 163–195.
37. Ishikawa, Y., & Tokoro, M., (1986). A concurrent object-oriented knowledge representation language orient84/K: Its features and implementation. *ACM SIGPLAN Notices, 21*(11), 232–241.
38. Jannarone, R. J., (1993). Concurrent information processing, I: An applications overview. *ACM SIGAPP Applied Computing Review, 1*(2), 1–6.
39. Jimack, P. K., & Topping, B. H. V., (1999). An overview of parallel dynamic load balancing for parallel adaptive computational mechanics codes. *Parallel and Distributed Processing for Computational Mechanics: Systems and Tools, 1*, 350–369.
40. Kan, G., He, X., Li, J., Ding, L., Hong, Y., Zhang, H., & Zhang, M., (2019). Computer aided numerical methods for hydrological model calibration: An overview and recent development. *Archives of Computational Methods in Engineering, 26*(1), 35–59.
41. Kasten, F. H., & Herrmann, C. S., (2019). Recovering brain dynamics during concurrent tACS-M/EEG: An overview of analysis approaches and their methodological and interpretational pitfalls. *Brain Topography, 32*(6), 1013–1019.
42. Kendall, R. A., Aprà, E., Bernholdt, D. E., Bylaska, E. J., Dupuis, M., Fann, G. I., & Wong, A. T., (2000). High performance computational chemistry: An overview of NWChem a distributed parallel application. *Computer Physics Communications, 128*(1, 2), 260–283.
43. Kirk, D., (2007). NVIDIA CUDA software and GPU parallel computing architecture. In: *ISMM* (Vol. 7, pp. 103–104).
44. Kshemkalyani, A. D., & Singhal, M., (2011). *Distributed Computing: Principles, Algorithms, and Systems* (Vol. 1, pp. 2–5). Cambridge University Press.
45. Ligon, III. W. B., & Ross, R. B., (1999). An overview of the parallel virtual file system. In: *Proceedings of the 1999 Extreme Linux Workshop* (Vol. 1, pp. 1–9).
46. Martins, R., Manquinho, V., & Lynce, I., (2012). An overview of parallel SAT solving. *Constraints, 17*(3), 304–347.

47. McBryan, O. A., (1994). An overview of message passing environments. *Parallel Computing, 20*(4), 417–444.
48. Motta, G., Sfondrini, N., & Sacco, D., (2012). Cloud computing: An architectural and technological overview. In: *2012 International Joint Conference on Service Sciences* (Vol. 1, pp. 23–27). IEEE.
49. Oduro, P., Nguyen, H. V., & Nieber, J. L., (1997). *Parallel Computing Applications in Groundwater Flow: An Overview* (Vol. 1, pp. 1–9). Paper-American Society of Agricultural Engineers.
50. Ohbuchi, R., (1985). Overview of parallel processing research in Japan. *Parallel Computing, 2*(3), 219–228.
51. Ostrouchov, G., (1987). Parallel computing on a hypercube: An overview of the architecture and some applications. In: *Computer Science and Statistics, Proceedings of the 19th Symposium on the Interface* (Vol. 1, pp. 27–32).
52. Paprzycki, M., & Zalewski, J., (1997). Parallel computing in Ada: An overview and critique. *ACM SIGAda Ada Letters, 17*(2), 55–62.
53. Parameswaran, M., & Whinston, A. B., (2007). Social computing: An overview. *Communications of the Association for Information Systems, 19*(1), 37.
54. Parashar, M., & Hariri, S., (2004). Autonomic computing: An overview. In: *International Workshop on Unconventional Programming Paradigms* (Vol. 1, pp. 257–269). Springer, Berlin, Heidelberg.
55. Parashar, M., & Lee, C. A., (2005). Grid computing: Introduction and overview. *Proceedings of the IEEE, Special Issue on Grid Computing, 93*(3), 479–484.
56. Park, D. A., (2008). Concurrent programming in a nutshell. *Journal of Computing Sciences in Colleges, 23*(4), 51–57.
57. Park, G. J., Lee, T. H., Lee, K. H., & Hwang, K. H., (2006). Robust design: An overview. *AIAA Journal, 44*(1), 181–191.
58. Pomello, L., (1985). Some equivalence notions for concurrent systems. An overview. In: *European Workshop on Applications and Theory in Petri Nets* (Vol. 1, pp. 381–400). Springer, Berlin, Heidelberg.
59. Pursula, M., (1999). Simulation of traffic systems-an overview. *Journal of Geographic Information and Decision Analysis, 3*(1), 1–8.
60. Qian, L., Luo, Z., Du, Y., & Guo, L., (2009). Cloud computing: An overview. In: *IEEE International Conference on Cloud Computing* (Vol. 1, pp. 626–631). Springer, Berlin, Heidelberg.

61. Ranganath, V. P., & Hatcliff, J., (2006). An overview of the Indus framework for analysis and slicing of concurrent java software (keynote talk-extended abstract). In: *2006 Sixth IEEE International Workshop on Source Code Analysis and Manipulation* (Vol. 1, pp. 3–7). IEEE.
62. Raynal, M., (2015). Parallel computing vs. distributed computing: A great confusion?(position paper). In: *European Conference on Parallel Processing* (Vol. 1, pp. 41–53). Springer, Cham.
63. Raynal, M., (2016). A look at basics of distributed computing. In: *2016 IEEE 36th International Conference on Distributed Computing Systems (ICDCS)* (Vol. 1, pp. 1–11). IEEE.
64. Saidu, C. I., Obiniyi, A. A., & Ogedebe, P. O., (2015). Overview of trends leading to parallel computing and parallel programming. *British Journal of Mathematics & Computer Science, 7*(1), 40.
65. Sérot, J., & Ginhac, D., (2002). Skeletons for parallel image processing: An overview of the skipper project. *Parallel Computing, 28*(12), 1685–1708.
66. Seymour, K., Nakada, H., Matsuoka, S., Dongarra, J., Lee, C., & Casanova, H., (2002). Overview of GridRPC: A remote procedure call API for grid computing. In: *International Workshop on Grid Computing* (Vol. 1, pp. 274–278). Springer, Berlin, Heidelberg.
67. Shrivastava, S. K., Dixon, G. N., & Parrington, G. D., (1989). An overview of Arjuna: A programming system for reliable distributed computing. *Computing Laboratory Technical Report Series, 1*, 4–8.
68. Shrivastava, S. K., Dixon, G. N., Hedayati, F., Parrington, G. D., & Wheater, S. M., (1988). A technical overview of Arjuna: A system for reliable distributed computing. *Computing Laboratory Technical Report Series, 1*, 3–9.
69. Singh, M., (2019). An overview of grid computing. In: *2019 International Conference on Computing, Communication, and Intelligent Systems (ICCCIS)* (Vol. 1, pp. 194–198). IEEE.
70. Stoughton, J. W., & Miekle, R. R., (1988). The ATAMM procedure model for concurrent processing of large grained control and signal processing algorithms. In: *Proceedings of the IEEE 1988 National Aerospace and Electronics Conference* (Vol. 1, pp. 121–128). IEEE.
71. Sunderam, V. S., Geist, G. A., Dongarra, J., & Manchek, R., (1994). The PVM concurrent computing system: Evolution, experiences, and trends. *Parallel Computing, 20*(4), 531–545.

72. Tasora, A., Negrut, D., & Anitescu, M., (2011). GPU-based parallel computing for the simulation of complex multibody systems with unilateral and bilateral constraints: An overview. *Multibody Dynamics, 1*, 283–307.
73. Teich, J., Henkel, J., Herkersdorf, A., Schmitt-Landsiedel, D., Schröder-Preikschat, W., & Snelting, G., (2011). Invasive computing: An overview. *Multiprocessor System-on-Chip, 1*, 241–268.
74. Thanisch, P., (2000). Atomic commit in concurrent computing. *IEEE Concurrency, 8*(4), 34–41.
75. Treleaven, P. C., (1988). Parallel architecture overview. *Parallel Computing, 8*(1–3), 59–70.
76. Vajteršic, M., Zinterhof, P., & Trobec, R., (2009). Overview–parallel computing: Numerics, applications, and trends. *Parallel Computing, 1*, 1–42.
77. Valentini, G. L., Lassonde, W., Khan, S. U., Min-Allah, N., Madani, S. A., Li, J., & Bouvry, P., (2013). An overview of energy efficiency techniques in cluster computing systems. *Cluster Computing, 16*(1), 3–15.
78. Walker, D. W., (1994). The design of a standard message passing interface for distributed memory concurrent computers. *Parallel Computing, 20*(4), 657–673.
79. Want, R., Schilit, B. N., Adams, N. I., Gold, R., Petersen, K., Goldberg, D., & Weiser, M., (1995). An overview of the PARCTAB ubiquitous computing experiment. *IEEE Personal Communications, 2*(6), 28–43.
80. Wu, Q., Spiryagin, M., & Cole, C., (2016). Longitudinal train dynamics: An overview. *Vehicle System Dynamics, 54*(12), 1688–1714.
81. Yang, X., Nazir, S., Khan, H. U., Shafiq, M., & Mukhtar, N., (2021). Parallel computing for efficient and intelligent industrial internet of health things: An overview. *Complexity, 1*, 5–9.
82. Yonezawa, A., (1994). Theory and practice of concurrent object-oriented computing. In: *International Symposium on Theoretical Aspects of Computer Software* (Vol. 1, pp. 365). Springer, Berlin, Heidelberg.
83. Zaslavsky, A., & Tari, Z., (1998). Mobile computing: Overview and current status. *Journal of Research and Practice in Information Technology, 30*(2), 42–52.

84. Zhang, G., Shang, Z., Verlan, S., Martínez-Del-Amor, M. Á., Yuan, C., Valencia-Cabrera, L., & Pérez-Jiménez, M. J., (2020). An overview of hardware implementation of membrane computing models. *ACM Computing Surveys (CSUR), 53*(4), 1–38.

CHAPTER 2

EVOLUTION OF CONCURRENT, PARALLEL, AND DISTRIBUTED COMPUTING

CONTENTS

2.1. INTRODUCTION

In the past, computers did not have access to system resources; instead, they would run a single program from start to finish. That program would have unrestricted access to those resources which the computer possessed. It was difficult to build programs which ran just on metal surfaces because running just a single application at a time was a wasteful use of costly and limited computational resources (Sunderam and Geist, 1999). Writing programs that operated just on metal surfaces was challenging. Operating systems have progressed to the point where they can now handle the execution of multiple programs at the same time. This is accomplished by running specific programs within processes, which are self-contained, and individually operated properly to which the system software assigns memory resources, file handles, and protection credentials (Tan et al., 2003). Processes have access to a wide array of poorly graded communication methods like connections, signal processors, memory units, synchronization primitives, and files, which they may use to interact with one another if the need arises. Several driving forces were responsible for the creation of computer systems that made it possible for numerous programs to run at the same time, including the following (Sunderam et al., 1994):

- **Resource Utilization:** It is occasionally necessary for programs to wait for other activities, including input or output because while they are waiting, they are unable to do any beneficial work. It will be more productive to let a program operate during that waiting period rather than doing nothing.
- **Fairness:** There might be several users and applications competing for the machine's resources, each in its unique way. It is better to let programs share the computer by allowing for more fine-grained time-slicing as opposed to letting one application run until it is finished and then beginning another.
- **Convenience:** It can often be easier or even more attractive to start writing many programs which each perform a specific task and also have them organize with one another as required than it is to post a personal program that needs to perform all of the tasks. This is because writing multiple programs which each perform a specific task allows for more flexibility.

In the earliest extension services processes, each procedure was a simulated general nonlinear computer. This meant that it had a system memory that could store both programs and data, that it would execute instructions in a specific order according to semantic information of the computer language, and that it would interact with the outside globe via the operating system using a set of I/O data structures (Tagawa et al., 2010). There had been a "next command" that was precisely outlined for each instruction that was carried out, and control moved through a program under the guidelines established by the processor architecture. This sequential computing approach is followed by almost all of the commonly used programming languages of today. Throughout this pattern, the language's description explicitly states "what happens next" once a certain action has been carried out (Geist and Sunderam, 1993).

2.2. EVOLUTION OF CONCURRENT COMPUTING

In the realm of train lines, telegram offices, and other comparable innovations, the phrase "Concurrent Computing" existed long earlier than computers were conceived (Venkat et al., 2021). Notwithstanding not being directly related to the computational problems that have plagued the last several years, the early understanding of parallelism made significant progress in solving problems such as rail line procedure parallelization, harmonizing, and having to send multiple data streams via telegraphs, and similar activities in several other sectors that arose during the industrialization and demanded parallel processes and systems. Recognize a late-18th-century factory with workers and machinery at various stages and assembly plants cranking out finished goods (perhaps using poked card readers – the 1890 US census used a tallying machine able to operate on punch cards), and you can see how multi-threading is it would have been an important concern also at that time (Yokote and Tokoro, 1987).

An abstraction and theoretical model of computing is a deterministic finite automaton (or finite state machine: FSM). Simply expressed, it represents an abstraction machine that takes inputs in the form of a series of characters which can only be in one system at a time. A transition happens when a state changes back and forth in time, and it is dependent on the input sign received from the sequences. The abstraction machinery only understands the present state it is in from such a global viewpoint, and there is no idea of storage that can keep a list of prior states. A highly artificial example of an FSM containing three states – A, B, and C – can be seen here (Figure 2.1) (Tasoulis et al., 2004).

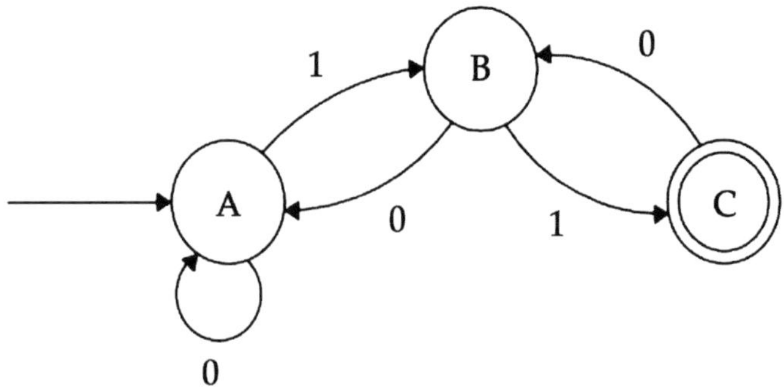

Figure 2.1. A, B, and C states of a finite state machine.

Source: https://markfaction.wordpress.com/2018/03/12/a-history-of-concurrent-programming-and-actor-based-frameworks-part-1/.

The beginning state is a state to which the first arrow points. State C, shown with the double circle, is the acceptable state. Transitional arrows connect the states and also self-refer to the state. The transition is labeled with symbolic meanings the input sign value that allows the transition to happen. Because there are three attributes and two input signals (0, 1) in the previous example, the FSM is referred to this as a three-state two-symbol FSM (Umemoto et al., 2004).

Given this knowledge, it is clear what will occur if an input pattern like 1011 is placed into FSM described above. Beginning state, A would be the very beginning state, from which we could extract the first sign of the input pattern 1011, which is 1. This causes the FSM to move from state A to state B, and now we are in state B. The next sign we read is 0, which indicates that we will return from B to A. Similarly, the following sign is 1, indicating that the FSM will move to state B, and the last symbol is 1, indicating that the state will move from B to C (Guirao et al., 2011; Ying, 2012).

Deterministic and non-deterministic FSMs (NFSM) are both possible. The FSM we just looked at is an instance of a determinism FSM, in which each state may only have one input transaction. An input in a NFSM may affect the current configuration of the FSM by providing one transition, several transitions, or no transformation at all. This introduces the concept of non-determinism to the FSM framework. An instance of a NFSM is shown below (NFSM) (Figure 2.2) (Poikolainen and Neri, 2013).

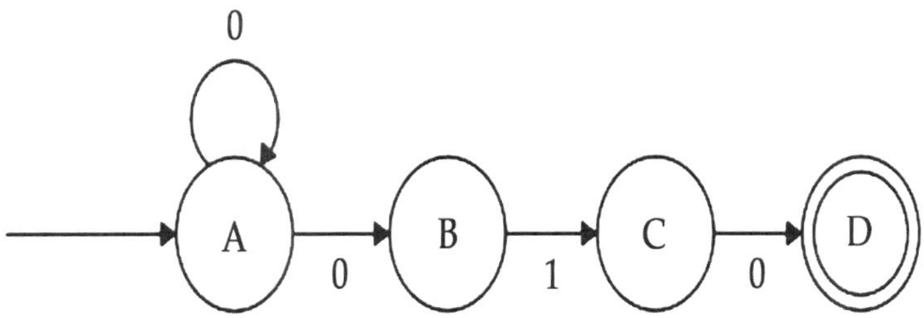

Figure 2.2. Finite state machine that is non-deterministic.

Source: https://markfaction.wordpress.com/2018/03/12/a-history-of-concurrent-programming-and-actor-based-frameworks-part-1/.

Another form of NFSM employs what is known as 'epsilon transition.' An epsilon transition, to put it another way, indicates that perhaps the state may change without using any input symbols. The following example of an NFSM using epsilon transitioning will help to clarify this (Figure 2.3) (Guan et al., 2002).

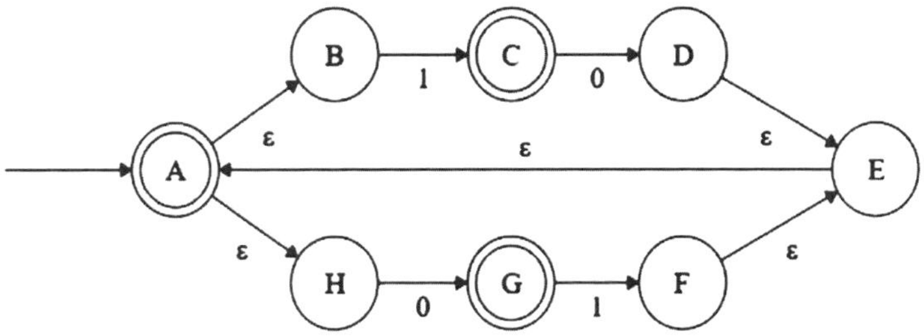

Figure 2.3. Epsilon transitions in NFSM.

Source: https://softwareengineering.stackexchange.com/questions/232471/epsilon-transitions-in-an-nfa.

The transition identified the with Greek letter epsilon () in the above NFSM are the epsilon transitions. State A is both the beginning point and an acceptable state. C and G are also acceptable states. Think about what happens if the sign 0 is the lone symbol in the input data set (symbol sequencing). First, we will go to state A, which is the beginning point. However, as there are two epsilon transfers from A, we shall proceed to the states indicated by all these without having to read in just about any input symbols. As a result, we immediately enter states B and H (Tagawa and Ishimizu, 2010). The input sign, which is 0, is already read in. Just in this scenario does state H shift to G, while state B does not change. However, since the input pattern is eaten and the NSFM ends in an acceptable state (G), we may deduce that 0 is part of the regular language that this NFSM accepts. As are 1, 01, 10, 010, 101, 1010, and so on (which can be checked against the above NFSM) (Mankel, 1997).

The strength of an embedded device is also another way of saying that it can identify more (ordinary) languages. FSMs have restrictions that are solved by a different sort of abstract device called pushdown automata (PDA) (Gill et al., 2019). This is a combination of an FSM and a stack that may be used to store data or status in a LIFO (Last in First Out) manner. This allows the PDA to keep data bits in the stacks but only come to an accepting state after the input sign sequence has been processed as well as the stacking is empty. This is a changed-over-time criterion than the FSMs we looked at before. A PDA in action is seen in Figure 2.4 (Janicki et al., 1986).

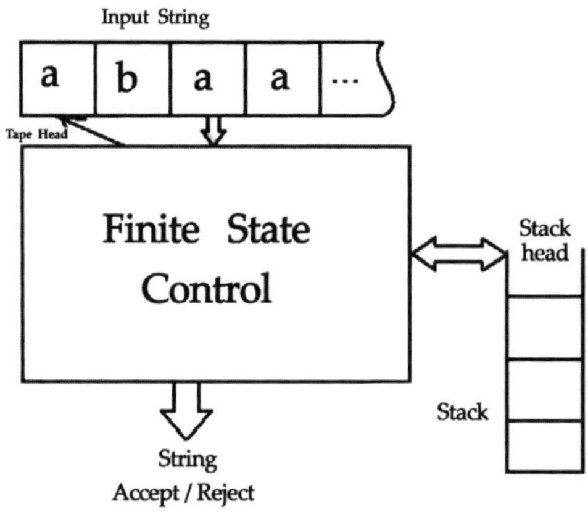

Figure 2.4. Automata that are pushed down (source).

Source: https://er.yuvayana.org/pushdown-automata-definition-formal-and-informal/.

If the strength of abstraction machines is measured by the languages they can understand, then Turing machines are more effective than either of FSMs or PDAs (This is an additional way of saying that there are orders of input symbols that cannot be treated by FSMs or PDAs, but which are putative by Turing machinery) (Xia et al., 2016).

Alan Turing initially introduced the Turing machine in 1936 in a work titled 'On Average Compared, with an Applications to the Factors may impact.' A Turing machine is an FSM with just an infinite supply of tape media to read English to (Rattner, 1985). In essence, the Turing machine gets to read a sign from the tape that serves as that of the possible input and then writes a sign to the recording, moves the tape left or right with one cell (a cell becomes one noticed unit of the tape that can consist a symbol), or transitions to a new state based on just this input symbol. This seems to be a simple and fundamental mode of operation, yet Turing devices are strong since we can build a Turing machine to replicate any method. Turing machines, furthermore, have the inherent potential to halt or stop, which leads to the stopping issue (Figure 2.5) (Cárdenas-Montes et al., 2013).

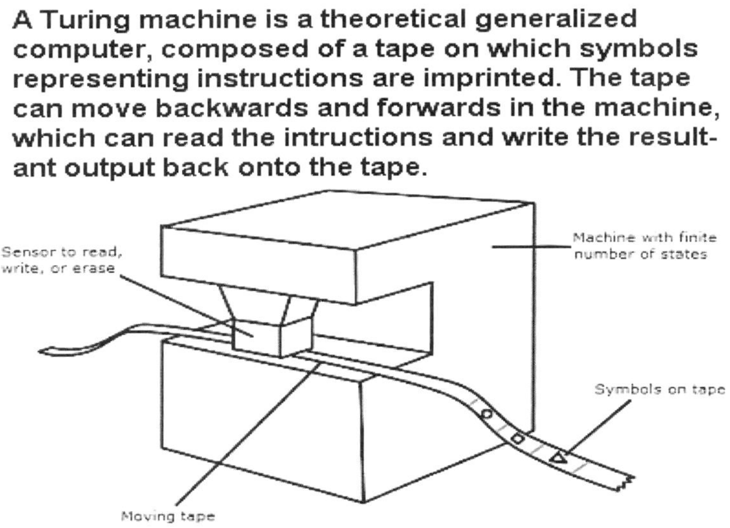

Figure 2.5. Model of a Turing machine in mechanical form.

Source: https://markfaction.wordpress.com/2018/03/12/a-history-of-concurrent-programming-and-actor-based-frameworks-part-1/.

If you want to explore Turing machines, go to this page, and go through the tutorial (I recommend starting with the 'equal number of zeros' example, which is simple but illustrates the basics). This movie also provides one of the greatest and most concise descriptions of Turing machines I have ever seen (Pelayo et al., 2010).

2.3. EVOLUTION OF PARALLEL COMPUTING

Parallel computing is based on the idea that a task may be broken down into smaller portions that can then be processed simultaneously. The idea of concurrency is connected to parallelism; however, the two phrases must not be confused. Concurrency is the construction of separate processes, whereas parallelism is the operation of several processes at the same time (Mühlenbein et al., 1988).

Since the late 1950s and early 1960s, capital-intensive sectors such as commercial aviation and defense have used distributed processing and similar ideas. With the significant decline in the cost of hardware over the last five decades and the emergence of open-source systems and apps, home hobbyists, students, and small businesses may now make use of these technologies for their purposes (Zhu, 2009).

2.3.1. Supercomputers

System Data Corporation, a business founded in the 1960s, was the forerunner of supercomputing (CDC). Seymour Cray was an electrical engineer at CDC who was dubbed the "Father of High-performance computing" for his research on the CDC 6600, widely regarded as the first powerhouse. Between 1964 and 1969, the CDC 6600 was the quickest computer in use (Skjellum et al., 1994).

Cray left CDC in 1972 and founded Cray Research, his firm. The Cray-1 supercomputer was unveiled in 1975 by Cray Research. The Cray-1 became one of the most productive supercomputers in humanity, and several institutions continued to utilize this until the late 1980s. Other companies entered the market in the 1980s, including Intel's Caltech Simultaneous Calculation project, which had 64 Intel 8086/8087 CPUs, and Thinking Machines Corporation's CM-1 Connections Machine (Crutchfield and Mitchell, 1995).

This was before a surge in the number of processing incorporated in high-performance computing computers in the 1990s. IBM notably defeated global chess champion, Garry Kasparov, with both the Deep Blue supercomputer during this decade, owing to brute-force processing power. Each of the 30 nodes of the Deep Blue computer had IBM RS6000/SP parallel processor and multiple "chess chips." By the early 2000s, there were tens of thousands of processors running in parallel (Omicini and Viroli, 2011). The Tianhe-2, which has 3,120,000 cores and can operate at 33.86 petaflops per second, held the position of best computer as of June 2013. Supercomputing is not the only place where parallel computing is used. These principles may now be found in multi-core and multi-processor desktop computers. We possess groups of autonomous devices, generally comprising a single core, that may be linked to function together across a network in addition to single devices. We will look into multi-core computers next since they are available at consumer technology stores all around the globe (Mühlenbein et al., 1987).

2.3.2. Multi-Core and Multiprocessor Machines

The use of machines with numerous cores and processors is no longer limited to supercomputing. However, how did we get to this stage if your laptop or smartphone has more than one processor core?

Moore's law has down the cost of elements, which has led to the widespread use of parallel processing. Moore's law states that every 18 to 24 months, the number of transistors in integrated circuits doubles. As a result, the hardware infrastructure such as CPUs has continually decreased. As a consequence, companies like Dell and Apple have developed even faster processors for the domestic market that easily exceed powerful computers that used to take up a whole room (Schmidt et al., 2002).

The 2013 Mac Pro, for example, has up to 12 cores, which means it has a CPU that replicates some of its major processing elements 12 times. These are a fraction of the cost of the Cray-1 when it was first released. We can experiment with parallel-based computing on a single system using devices with many cores (Vondrous et al., 2014). Threads are one approach for making use of numerous cores. Threads are a set of instructions that are normally encapsulated in a single compact process that the system software may plan to perform. From the standpoint of computing, this might be a distinct function that operates independently of the program's primary cores. Because of the ability to employ threads in program development, POSIX Threads (Pthreads) and OpenMP had grown to dominate the domain of distributed memory multiprocessing devices even by the 1990s (Yong et al., 1995).

2.3.3. Commodity Hardware Clusters

We have clusters of commodities off-the-shelf (COTS) computers that can be connected to a local area network (LAN), just as we have individual devices with numerous CPUs (LAN). Beowulf clusters were formerly a popular name for them. The construction of Beowulf clustering became a prominent subject in the late 1990s, due to a reduction in the cost of computer equipment, and Wired magazine published its first how-to guide in 2000: Beowulf is the term given to Donald J. Becker and Thomas Sterling's notion of a network of workstations (NOW) for computer science, which was developed at NASA in the early 1990s. The Raspberry P_i-based applications we will be constructing in this book are constructed based on commodity hardware clusters that run technology like MPI (Vrugt et al., 2006).

2.3.4. Cloud Computing

A group of technology that is dispersed, modular, metered (as with utilities), may be executed in parallel, and typically incorporate virtualization layers are now at the heart of the term. Virtually hardware is software that can be designed as though it were a genuine machine and replicates the function of a genuine hardware component. VirtualBox, Red Hat Enterprise Virtualization, and parallelism virtualized software are only some of the virtual machine software options (PVM) (Lastovetsky, 2008). More information about PVM may be found at: Many significant Internet-based organizations, including Amazon, have engaged in cloud technology during the last decade. Amazon built a cloud software architecture after realizing they were underusing a substantial chunk of its data centers, ultimately leading to the creation of Amazon web services (AWS), a public platform (Mühlenbein et al., 1991).

Small companies and home customers may now hire virtual computers to operate their apps and services thanks to products like Amazon's AWS elastic compute cloud (EC2), which allows users to rent virtual machines to operate their apps and services. This is particularly handy for individuals who want to create virtual computer clusters. It is simple to spool up several multiple servers and connect them just to explore with technology like Hadoop thanks to the flexibility of cloud computing platforms like EC2. The analysis of large data is one area wherein cloud technology has proved very useful, particularly when utilizing Hadoop (Mehlführer et al., 2009).

2.3.5. Big Data

The phrase "big data" now refers to data collections that are terabytes in size or more. Big data sets are tough to work with it and demand a lot of memory and computer capacity to query. They are common in industries like genetics and astronomy (Venkatasubramanian, 1999). The information must be extracted from these data sets. The use of parallel technology like MapReduce, as implemented in Apache Hadoop, has offered a method for partitioning a huge operation like this across numerous processors. After the jobs have been separated, the data must be located and compiled. Hive, a Hadoop data storage system that uses a SQL-like special language HiveQL to query the stored data, is another Apache project (Felder, 2008).

The capacity to manage massive datasets and analyze them simultaneously to speed up data searches will become increasingly vital as more computational resources, ranging from sensing to cameras, create more data each year. These big data issues have pushed the bounds of parallel computing even further, as several organizations have sprung up to assist in the extraction of information from the volume of information that now abounds (Migdalas et al., 2003).

2.4. EVOLUTION OF DISTRIBUTED COMPUTING

Thanks to the emergence of large network interconnections, many of the resources required by programming no longer remain on a single system. A bank card, for instance, might need to be verified on a company's computer, with the current computer's program alone responsible for passing the information to the company's computer. Another example is an application that interfaces with an existing system, including a large employees databases developed in COBOL and running on a supercomputer (Lee and Benford, 1995). An instance is any internet program that transmits impulses to a web application on a separate computer and receives a response. It is difficult to conceive that a developer nowadays could do their job without writing any software that uses networking computing. Because programs that run on separate machines must run at the same time, all of the ideas shown here will apply to distinct parts of the contemporaneous software system (Suciu et al., 2013).

It is helpful to understand how distributed computing evolved throughout time to better understand the current technology. Initially, dispersed systems are built via a "running shoe net," which consisted of rolling tapes transporting data files to many computers that executed the data processing algorithms. It was a progression over running shoe net while software engineers-built screenplays using standard water utilities, including FTP and UUCP, to recreate these file formats across both computer systems; even so, such various types of alternative solutions were packet-oriented, so documents were transferred between the procedures and put in place one or even more times a day (Kshemkalyani and Singhal, 2011).

At some point, a physical access to remote systems became required. This implies that, initially, developers had to design their protocol for communicating to faraway systems, which were centered on limited cavemen such as connections. While connections produced it simple to begin writing files between devices, they just perused and wrote simple bytes streams, leaving the software to deal with problems such as the range of data kinds between many computer systems (e.g., how to personify flying positions and massive vs. slight queries) and trying to rebuild the byte streams into the productive program provided. Actual dialog amongst programs was expensive to accomplish since the programer had to contend with negative Neanderthals and construct and develop their protocols for sending and receiving data (Balabanov et al., 2011).

RPC had one flaw: this only functioned with procedures, not items. Other technologies, like the Common Object Brokerage Architectural (CORBA) or the Config file simple object stipulated conditions (SOAP), were developed to make objects available across networking. Remote procedure invoking (RMI) is a Programming language mechanism for transferring and accessing artifacts via the internet. It makes use of the Core java language's features to create infrastructure items (Sunderam et al., 1994). We will follow the timetable in Figure 2.6 to see how things evolve.

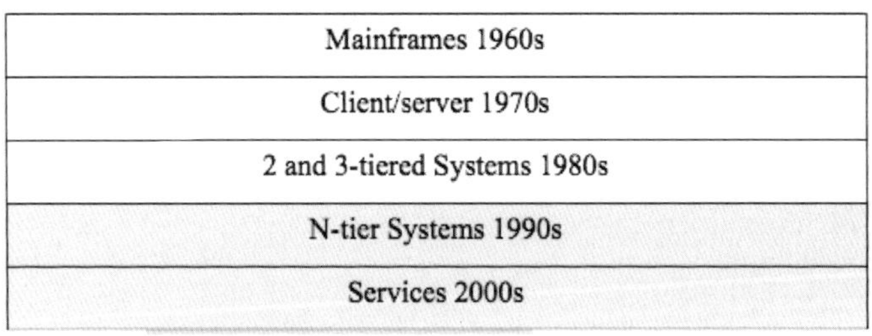

Figure 2.6. A timeline for distributed systems.

Source: https://userpages.umbc.edu/~jianwu/is651/651book/is651-strapdown.php?f=is651-Chapter02.md.

This chapter will describe all of these designs and the rest of the book will focus on N-tier networks and applications. The history does not precisely match the decades indicated in our timeline, but it gives us a decent idea of where the different periods are located (Joseph et al., 2004).

2.4.1. Mainframes (the 1960s)

Initially, a commercial system implied that a company could only have one supercomputer (CPU). Software on compact discs would be required to see the output from a printer (see Figure 2.7). A punctured card is indeed a stiff piece of paper that has digital information stored on it by including or excluding holes at predefined points (De Roure et al., 2003). It was a late-19th century advancement in textiles looms operation technology. One used a keypad to input the programming into a critical machine, causing holes in a deck of cards. After that, the cards are put into a memory stick processor, that analyzed these and communicated the software to the computer for processing. This has been processed, and the outcome was generated after a wait while one's work sat in a systems queue. A computer programer would generally take that print and deposit it in a postal cubby from whence this could be retrieved (Megino et al., 2017).

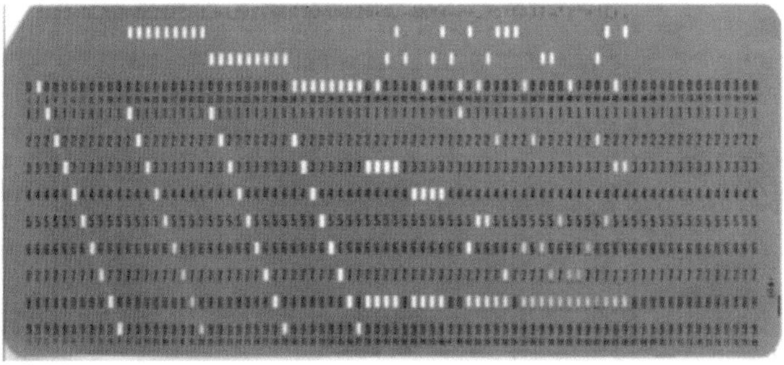

Figure 2.7. A punch card is a piece of paper with a number on it.

Source: https://www.extremetech.com/computing/90156-the-history-of-computer-storage-slideshow/2.

In the 1960s, ray tube (CRT connectors were added, enabling users to program the computer and then see the output on a display, which was significantly more efficient. Furthermore, because there was only one CPU (or numerous CPUs inside one box), the notion of a distributed system still was rather simple. Because they lacked a CPU, the interfaces used only for access were dubbed "stupid" interfaces (Jia et al., 2018).

2.4.2. Client/Server Computing (the 1970s)

Within the 1970s, microcomputers, the Web, or X.25 have all invented. Microprocessors were tiny processors that could be employed in a larger number of organizations. Internet established the distinction between local-area (LAN) and wide-area (WAN) communication (WAN). The LAN enabled microcomputers to communicate as servers and clients via Internet, something you learnt about that in your basic networking course (Mikkilineni et al., 2012). A consumer is changing order procedures (not a computer) and a server is a response technique (remember, not a device or server), and should both have CPUs in such a 2-tier design. The X.25 method was developed as a Network architecture that was the first packet-switching technology for digital information (Figure 2.8) (Satyanarayanan, 1996).

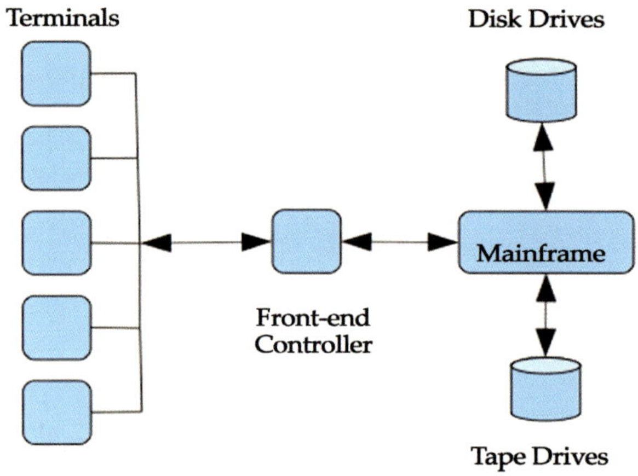

Figure 2.8. Architecture for mainframe computers.

Source: https://userpages.umbc.edu/~jianwu/is651/651book/is651-strapdown.php?f=is651-Chapter02.md.

The Internet was born in the 1970s, with the introduction of the advanced research projects agency network in 1969, which connected computers primarily at universities and Ministry of Defense installations. The TCP/IP framework was established in 1978 (Desell, 2017).

2.4.3. 2 and 3-Tier Systems (the 1980s)

The first real and pervasive scattered networks arose in the 1980s. The IBM PC, which first appeared in 1981, was revolutionary in that it enabled everyone to possess a computer and link it to others through a LAN. Novell's NetWare has become the most frequently utilized LAN technology platform, delivering file servers within the first big deployment of a client/server architecture integrating IBM PC consumers operating MS-DOS and Network equipment storage space (Bencina and Kaltenbrunner, 2005). The very first network diagram, the networking file system (NFS), was built in 1985 for Linux operating system minicomputers. We will examine different filetypes throughout this book. TCP/IP was originally used to connect smaller networks to the Advanced research projects agency network in 1982. The domain controller (DNS) was founded in 1984 and is still used to identify people online today. In 1989, Timothy Berners-Lee established HTTP, which would be the web's basis (WWW) (Li and Zhou, 2011).

Bridging and gateways, which allowed for the construction of networks with a more complicated topology, were more popular during the late 1980s. The website and accompanying elements served as an architectural intermediary tier between the client and server as a consequence of this design, resulting in true 3-tier computing (Dhar, 2012; Baldoni et al., 2003). Middleware is computing software that runs "in the middle," offering services to distributed platforms on the second through Nth layers. It is what gives the impression of a unified system to a customer. As we have already seen later, the software is not required to be stored in a tier, but it is often distributed throughout them to give the very same abilities and interactions to all nodes. For further details, please see Figure 2.5. The major types of network elements are as follows (Figure 2.9) (Taylor et al., 2015):

- Remote method calls need software (RPC);
- Message-oriented middleware (MOM) is a type of middleware which is used to send and receive messages.

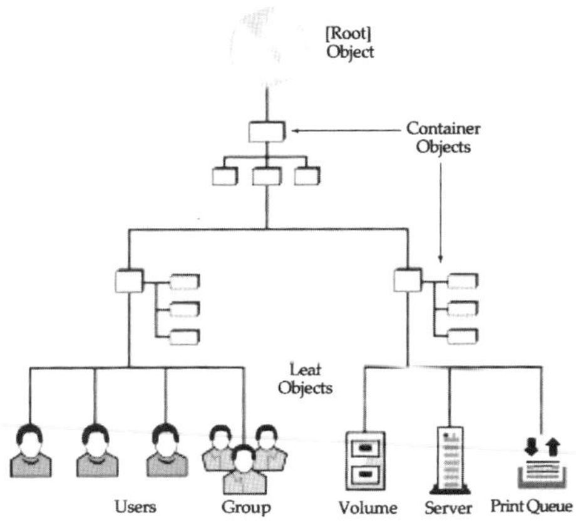

Figure 2.9. A NetWare network is a network that uses the NetWare operating system.

Source: https://www.wisdomjobs.com/e-university/networking-tutorial-273/novell-netware-31.html.

2.4.4. RPC Middleware

RPC (remote procedure call) is a gateways program that allows people to make requests to processes on cloud computers as if they were individual part operations to a software engineer (Figure 2.10). This is accomplished using the RPC gateways software design, which eliminates the need for specific compilers. The developer only builds programs that contact operations, whereas the RPC gateways handle remote calls. Figure 2.11 depicts the RPC's basic architecture (Chong and Kumar, 2003).

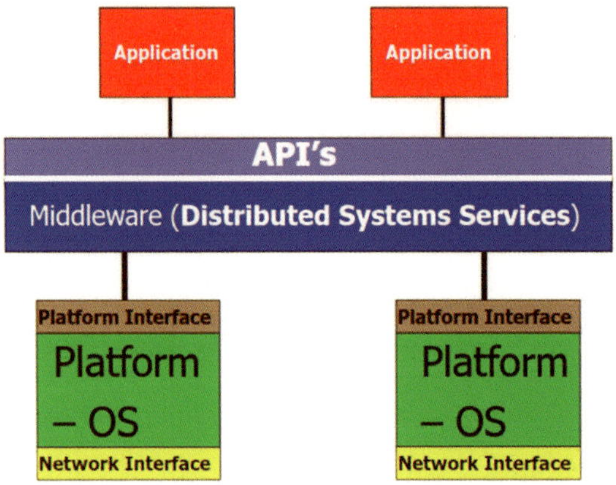

Figure 2.10. Middleware.

Source: https://medium.com/@rashmishehana_48965/middleware-technologies-e5def95da4e.

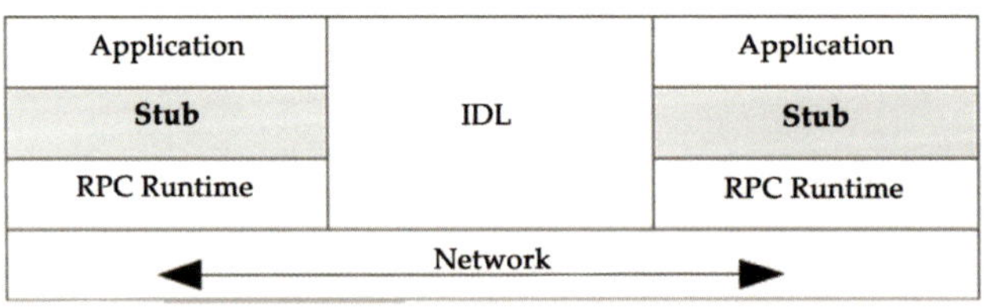

Figure 2.11. RPC (Remote procedure call) architecture.

Source: https://userpages.umbc.edu/~jianwu/is651/651book/is651-strapdown.php?f=is651-Chapter02.md.

Distributed data systems, like the network file system (NFS), discussed above, are a sort of RPC middleware. Users may mount directories subtrees from faraway machines into their local specified directory using the same kind of software. Although it seems to the client that all files and data are local, this is not the reality (Di Stefano, 2005).

Figure 2.12 shows how database model 1 may mount hierarchies' subsequences from file system 2 using the NFS protocol. NFS is a compendium of procedures, which include XDR, RPC, and NFS, that work at different stages of both the OSI model. RPC handles networking communication, XDR handles the mechanics of converting incompatible file types from multiple operating systems, and NFS handles file system accessibility. Take notice of the habit of identifying protocols layering by the identification of just one (or very few) of the conventions (Westerlund and Kratzke, 2018).

OSI Layer	File System 1	File System 2
Application	NFS	NFS
Presentation	XDR	XDR
Session	RPC	RPC
Transport	TCP	TCP

Figure 2.12. The NFS file system is a distributed file system.

In a TPM, the concept of a two-phase commitment plays a critical role (Figure 2.13). Inside this case, the steps of a procedure were not just numerous but also dispersed throughout several servers. The client makes a user request to the intermediate TPM host computer, which further requests that each of the services does the required actions (De et al., 2015; Feinerman and Korman, 2013). The TPM maintains track of the overall transaction results, enabling it to publish the entire transaction or dial it back when one of the components fails. Each server accomplishes one's own set of tasks and reports to the TPM whether this is published or moves back. After completing or scaling back the transaction, the TPM wait until all of the customers have responded, thus there are two stages: (i) each user's implementation of the semi; and (ii) the TPM's assessment of the total transaction. It is worth mentioning that the simultaneous functionality of RPC is very useful for transactions since we do not want delays (Huh and Seo, 2019).

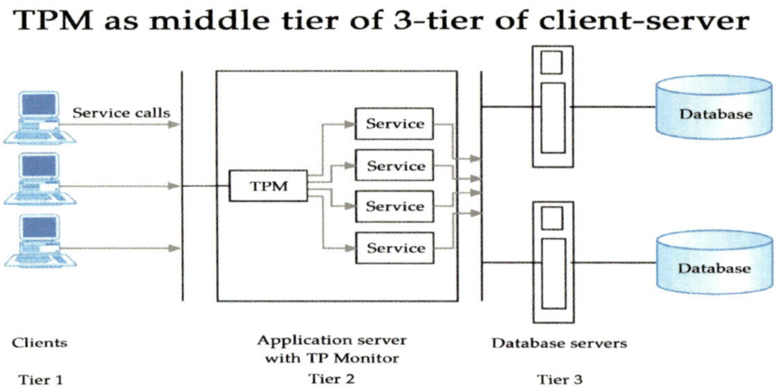

Figure 2.13. Total productive maintenance (TPM).

Source: https://slidetodoc.com/1-chapter-2-database-environment-2-objectives-purpose/.

Entity principles provided the foundation for RPC-based networks in the 1980s, while object-oriented computers became popular. Maybe the most well-known was the CORBA (comes mainly demand broker architectural) technology. According to with technology concept, a query language broker (ORB) relayed a callback from one property to another within a network. All objects should be recorded by the ORB. Figure 2.14 depicts CORBA's design (Lee, 2013).

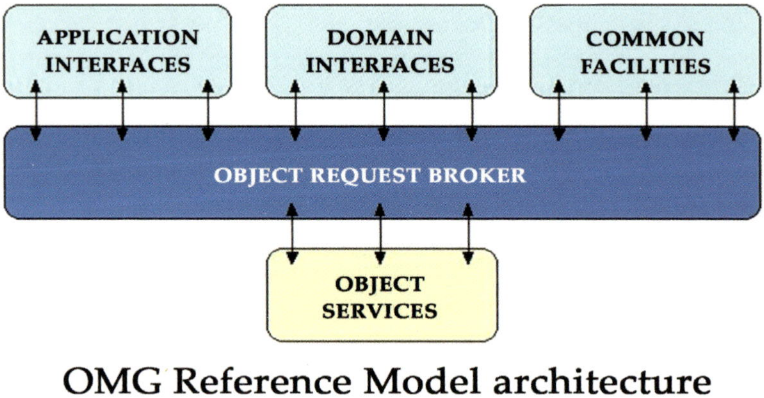

Figure 2.14. CORBA.

Source: https://networkencyclopedia.com/common-object-request-broker-architecture-corba/.

Procedures designed to monitor may be invoked by consumer entities. It is equivalent to regular RPC since the client has a stub. Although the user's object has a skeleton, it is just a CORBA naming quirk; it also is a stub. Also, it is worth mentioning that different ORBs may communicate with each other via the TCP/IP stack's communication mechanism. To use the internet Inter-ORB protocol, ORBs from diverse providers may communicate over the Web (IIOP) (Padhy et al., 2012).

2.4.5. Message-Oriented Middleware (MOM)

Message-oriented mediation (MOM) sends messages between computers in a dispersed system, as even the name suggests. As a consequence of this packet transmission, MOM may be able to manage synchronized communications. This means that, like RPC, a server may send out a message that includes an order to run the application, however, the host application would not have to wait for a response. Whenever the process gets the response, it simply continues its operation after a little stop caused by the returned messages (Andrews and Olsson, 1986).

This synchronous nature works well in a variety of settings. It gives you more flexibility since it does not need a quick response. For example, a host is not always available to receive messages in its route, but if it is, it may deliver the queue times content. This is not, though, appropriate for all purposes. It is more complicated to create and is not appropriate for time-critical operations like payments. MOM is a less tightly coupled design paradigm than RPC so it does not need as much integration (Böhm et al., 2010).

MOMs come in two varieties (Schmitt et al., 2015):

- Messaging on a one-to-one basis (PTP); and
- Publish/Subscribe messaging (pub/sub) is a kind of communication that allows you to send and receive messages.

PTP communication structure is shown in Figure 2.15. Thc main distinction was that by being transmitted instantly to the intended receiver, messages are routed to a backlog. Queuing is just a collection of information that is stored until the processing system is ready to process them. They can be delivered in any order, including "first come, first served," "preference," and "initially come, first delivered" (Shih, 2001).

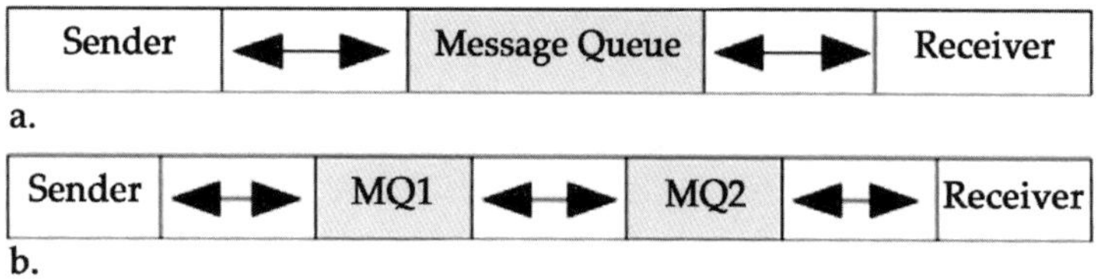

Figure 2.15. Message queues are a kind of message queue (MQ).

Source: https://userpages.umbc.edu/~jianwu/is651/651book/is651-strapdown.php?f=is651-Chapter02.md.

In Figure 2.15, the first scenario (a) involves message exchange over a shared MQ. The second scenario (b) involves each host keeping its queue. The queue central processing unit (CPU) has the capacity to send information to other queues. MOM is used in the industrial world by IBM's MQSeries and WebSphere messaging broker (WMB) (Ganek, 2004). An instance is MSMQ from Microsoft. MOM's apps include the Java mobile-based social (JMS). One of MOM's flaws is that there are no rules. Because there is no global protocol or communications format, each provider or programer executes it in their unique way. A developing standardization for MOM is the enhanced messaging protocol (AMQP), which defines a protocol and message format that allows different versions to interact with one another (Satyanarayanan, 1993).

PubSubHubbub is an accessible pub/sub implementation that extends the Atomic and very basic syndicated (RSS) protocols for sources of data. A subscribe fetches a feed as normal while also subscribing towards the hub publishing server if desired. When a feed is updated, the host may notify or push the subscription (Fritzsche et al., 2012).

2.4.6. N-Tier Systems (the 1990s)

Communications proliferated at a dizzying speed in the 1990s. The quantity of broadband accessible has increased dramatically. For example, LANs have progressed from 10 megabits to 100 mbps to 1,000 megabits per second. The (Important research programs agency connection) was pulled out in favor of the publicly available Internet in 1990, and the first internet providers (ISPs: internet service providers) sprung up (Gusev and

Dustdar, 2018). The popularity of the Internet skyrocketed, and it had a tremendous influence on the expansion of software systems. Even before Net, WANs were extremely expensive. Individual lines or services would also have to be rented from better-serve customers, which would have been prohibitively expensive for all but the largest businesses. Because of the Web, everyone had WAN connectivity. HTTP, HTML, and URLs/DNS were utilized to promote the WWW, which proved to be a major success for both corporations and individuals (Stahl and Whinston, 1994). This pairing spawned an e-commerce craze that continues even now. Unicode was established in 1996 and is currently the Web's norm today, allowing for 16-bit symbols. During this time, transnational naming systems were established or extensively utilized, notably the dos attack (DNS) for such Web with NetWare file systems (NDS) for local networks. Microsoft has risen to the top of the market for hardware and software, office software, and networking. Java is a widely-used programming language (Figure 2.16) (Milenkovic et al., 2003).

Presentation - Web Browser (client)
Communication - Web Server
Logic - Application Server
Storage - Database Server

Figure 2.16. The four-tier architecture for the web.

Source: https://userpages.umbc.edu/~jianwu/is651/651book/is651-strapdown.php?f=is651-Chapter02.md.

A web browser, for example, is a very thin consumer and a separate software program, similar to how a physical interpretation is. Its goal is to give information to the user, with the logic knowledge serving as a session layer. The physical web service, on the other hand, is responsible for mediating requests from browsers to the host computer, hence the season for example logical word is telecommunications (Figure 2.17) (Bąk et al., 2011).

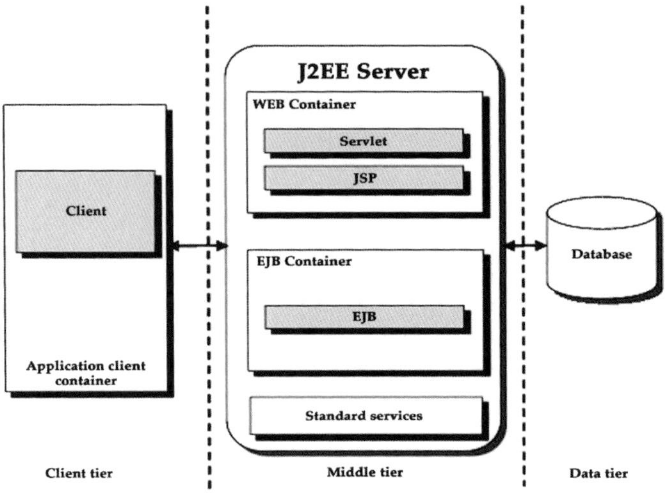

Figure 2.17. In J2EE, an EJB container is used.

Source: https://www.researchgate.net/figure/The-J2EE-Software-Architecture_fig1_220817684.

2.4.7. Services (the 2000s)

The goal of the 2000s was to implement provider architecture (SOA: service-oriented architecture). Instead of a set of technology, SOA is a portfolio of design principles. The following is a quick rundown of these fundamentals (Mikkilineni and Sarathy, 2009):

- **Standard Service Contracts:** The participants provided standard service contracts. They should be trackable via a registry or directories. We will look at how this works with XML web applications.
- **Loose Coupling:** We have already established the importance of individuals having a minimal dependency on each other.
- **Encapsulation:** Companies should keep their rationale hidden from the outside community as if they were a black box. This increases the company's reuse and customization (combining services). Consumers should not be concerned about where solutions are housed, hence services should be geographically transparent.

- **Statelessness:** Keep as little condition in attention as possible. This is a requirement when it comes to direct connection and encapsulation.

The entire book focuses on XML internet providers, which are used to build SOA. That is a Config file technical framework that standardizes XML interface across a variety of systems that may be used in any application framework, including J2EE.NET, and programming. These approaches are usually given over for an HTTP web communication and are detached from the implementations. The three main XML wire techniques for web services are SOAP, WSDL, and UDDI. The inquiry asker uses the UDDI protocol to find the services it needs, then uses the WSDL method to figure out where to communicate and then use it. The SOAP method is used to make the real request. Figure 2.18 depicts the simplified architecture of XML online services (El-Sayed et al., 2017).

2.4.8. Markup Languages

XML, like HTML, is a markup language. We have to learn how markup language functions to comprehend the technologies at an adequate technical level (Brugali and Fayad, 2002).

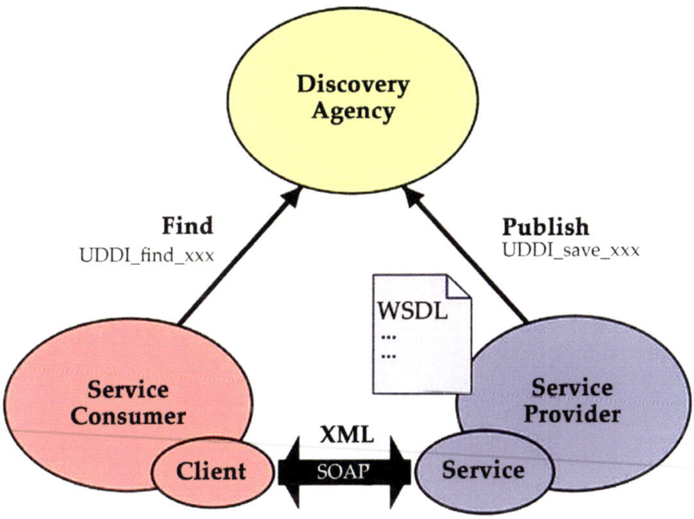

Figure 2.18. Architecture for XML web services.

Source: https://www.flickr.com/photos/dullhunk/415645479.

In 1989, Tim Berners-Lee designed HTML, as we are all familiar with thanks to websites. IBM researchers formulated the standardized general programming language (SGML) in the 1960s, which he utilized to create HTML. It was built on the very same foundation as today's modern computer languages. The markup is made up of labels or elements that are separate from the content of the page. The most common tag is vocabulary, whereas the element is a much more formal one. They always refer to the same thing but may be utilized in any situation. Here is a basic instance (Kaur and Kaur, 2014):

<section>Section I</section>

As a consequence, Tim Berners-Lee created a DTD again for internet language HTML built on SGML. HTML was designed to construct websites in mind, and practically all of its properties are utilized for that purpose. The relevant text is bolded in the HTML code following, for instance (Rimal et al., 2009).

<strong>the brown fox</strong>

SGML was indeed a specialized and obscure topic before HTML. As the web increased in popularity and also the Web extended substantially, HTML had become more useful for a distributed system. As B2B activity just on the net grew in prominence, users were dissatisfied with both the SGML foundation. SGML was created in the 1960s and was not intended for use on the internet (Dede, 1996). As a consequence, the XML 1.0 standards have been developed in 1998 as an internet successor for SGML, removing numerous of the SGML elements that are no longer relevant or useful. HTML has indeed been recast as an XHTML DTD and now is recognized as XHTML, which is a conceptual akin to SGML. In XML, DTDs are kept for consistency, and XMLSchema, a more modern schema extensible markup, has already been developed. This chapter will explore the topics of XML and DTD authenticity (Dai and Li, 2007).

REFERENCES

1. Andrews, G. R., & Olsson, R. A., (1986). The evolution of the SR language. *Distributed Computing, 1*(3), 133–149.
2. Bąk, S., Krystek, M., Kurowski, K., Oleksiak, A., Piątek, W., & Wąglarz, J., (2011). GSSIM–a tool for distributed computing experiments. *Scientific Programming, 19*(4), 231–251.
3. Balabanov, T., Zankinski, I., & Dobrinkova, N., (2011). Time series prediction by artificial neural networks and differential evolution in distributed environment. In: *International Conference on Large-Scale Scientific Computing* (Vol. 1, pp. 198–205). Springer, Berlin, Heidelberg.
4. Baldoni, R., Contenti, M., & Virgillito, A., (2003). The evolution of publish/subscribe communication systems. In: *Future Directions in Distributed Computing* (Vol. 1, pp. 137–141). Springer, Berlin, Heidelberg.
5. Bencina, R., & Kaltenbrunner, M., (2005). The design and evolution of fiducials for the reactivision system. In: *Proceedings of the Third International Conference on Generative Systems in the Electronic Arts* (Vol. 2, pp. 1–6). Melbourne, VIC, Australia: Monash University Publishing.
6. Böhm, M., Leimeister, S., Riedl, C. H. R. I. S. T. O. P. H., & Krcmar, H., (2010). *Cloud Computing and Computing Evolution* (Vol. 1, pp. 2–8). Technische Universität München (TUM), Germany.
7. Brugali, D., & Fayad, M. E., (2002). Distributed computing in robotics and automation. *IEEE Transactions on Robotics and Automation, 18*(4), 409–420.
8. Cárdenas-Montes, M., Vega-Rodríguez, M. Á., Sevilla, I., Ponce, R., Rodríguez-Vázquez, J. J., & Sánchez, Á. E., (2013). Concurrent CPU-GPU code optimization: The two-point angular correlation function as case study. In: *Conference of the Spanish Association for Artificial Intelligence* (Vol. 1, pp. 209–218). Springer, Berlin, Heidelberg.
9. Chong, C. Y., & Kumar, S. P., (2003). Sensor networks: Evolution, opportunities, and challenges. *Proceedings of the IEEE, 91*(8), 1247–1256.
10. Crutchfield, J. P., & Mitchell, M., (1995). The evolution of emergent computation. *Proceedings of the National Academy of Sciences, 92*(23), 10742–10746.

11. Dai, F., & Li, T., (2007). Tailoring software evolution process. In: *Eighth ACIS International Conference on Software Engineering, Artificial Intelligence, Networking, and Parallel/Distributed Computing (SNPD 2007)* (Vol. 2, pp. 782–787). IEEE.
12. De Roure, D., Baker, M. A., Jennings, N. R., & Shadbolt, N. R., (2003). *The Evolution of the Grid. Grid Computing: Making the Global Infrastructure a Reality, 2003*, 65–100.
13. De, K., Klimentov, A., Maeno, T., Nilsson, P., Oleynik, D., Panitkin, S., & ATLAS Collaboration, (2015). The future of PanDA in ATLAS distributed computing. In: *Journal of Physics: Conference Series* (Vol. 664, No. 6, p. 062035). IOP Publishing.
14. Dede, C., (1996). The evolution of distance education: Emerging technologies and distributed learning. *American Journal of Distance Education, 10*(2), 4–36.
15. Desell, T., (2017). Large scale evolution of convolutional neural networks using volunteer computing. In: *Proceedings of the Genetic and Evolutionary Computation Conference Companion* (Vol. 1, pp. 127, 128).
16. Dhar, S., (2012). From outsourcing to cloud computing: Evolution of IT services. *Management Research Review, 35*(8), 664–675.
17. Di Stefano, M., (2005). *Distributed Data Management for Grid Computing* (Vol. 1, pp. 2–6). John Wiley & Sons.
18. El-Sayed, H., Sankar, S., Prasad, M., Puthal, D., Gupta, A., Mohanty, M., & Lin, C. T., (2017). Edge of things: The big picture on the integration of edge, IoT and the cloud in a distributed computing environment. *IEEE Access, 6*, 1706–1717.
19. Feinerman, O., & Korman, A., (2013). Theoretical distributed computing meets biology: A review. In: *International Conference on Distributed Computing and Internet Technology* (Vol. 1, pp. 1–18). Springer, Berlin, Heidelberg.
20. Felder, G., (2008). CLUSTEREASY: A program for lattice simulations of scalar fields in an expanding universe on parallel computing clusters. *Computer Physics Communications, 179*(8), 604–606.
21. Fritzsche, M., Kittel, K., Blankenburg, A., & Vajna, S., (2012). Multidisciplinary design optimization of a recurve bow based on applications of the autogenetic design theory and distributed computing. *Enterprise Information Systems, 6*(3), 329–343.

22. Ganek, A., (2004). Overview of autonomic computing: Origins, evolution, direction. In: Salim, H. M. P., (ed.), *Autonomic Computing: Concepts, Infrastructure, and Applications* (Vol. 1, pp. 3–18).
23. Geist, G. A., & Sunderam, V. S., (1993). The evolution of the PVM concurrent computing system. In: *Digest of Papers; Compcon. Spring* (Vol. 1, pp. 549–557). IEEE.
24. Gill, S. S., Tuli, S., Xu, M., Singh, I., Singh, K. V., Lindsay, D., & Garraghan, P., (2019). Transformative effects of IoT, blockchain and artificial intelligence on cloud computing: Evolution, vision, trends and open challenges. *Internet of Things, 8*, 10.
25. Guan, Q., Liu, J. H., & Zhong, Y. F., (2002). A concurrent hierarchical evolution approach to assembly process planning. *International Journal of Production Research, 40*(14), 3357–3374.
26. Guirao, J. L., Pelayo, F. L., & Valverde, J. C., (2011). Modeling the dynamics of concurrent computing systems. *Computers & Mathematics with Applications, 61*(5), 1402–1406.
27. Gusev, M., & Dustdar, S., (2018). Going back to the roots—The evolution of edge computing, an IoT perspective. *IEEE Internet Computing, 22*(2), 5–15.
28. Huh, J. H., & Seo, Y. S., (2019). Understanding edge computing: Engineering evolution with artificial intelligence. *IEEE Access, 7*(2), 164229–164245.
29. Janicki, R., Lauer, P. E., Koutny, M., & Devillers, R., (1986). Concurrent and maximally concurrent evolution of nonsequential systems. *Theoretical Computer Science, 43*(1), 213–238.
30. Jia, Y. H., Chen, W. N., Gu, T., Zhang, H., Yuan, H. Q., Kwong, S., & Zhang, J., (2018). Distributed cooperative co-evolution with adaptive computing resource allocation for large scale optimization. *IEEE Transactions on Evolutionary Computation, 23*(2), 188–202.
31. Joseph, J., Ernest, M., & Fellenstein, C., (2004). Evolution of grid computing architecture and grid adoption models. *IBM Systems Journal, 43*(4), 624–645.
32. Kaur, R., & Kaur, A., (2014). A review paper on evolution of cloud computing, its approaches and comparison with grid computing. *International Journal of Computer Science and Information Technologies, 5*(5), 6060–6063.

33. Kshemkalyani, A. D., & Singhal, M., (2011). *Distributed Computing: Principles, Algorithms, and Systems* (Vol. 1, pp. 2–9). Cambridge University Press.
34. Lastovetsky, A. L., (2008). *Parallel Computing on Heterogeneous Networks* (Vol. 1, pp. 1–5). John Wiley & Sons.
35. Lee, J., (2013). A view of cloud computing. *International Journal of Networked and Distributed Computing, 1*(1), 2–8.
36. Lee, O. K., & Benford, S., (1995). An explorative model for federated trading in distributed computing environments. In: *Open Distributed Processing* (Vol. 1, pp. 197–207). Springer, Boston, MA.
37. Li, Q., & Zhou, M., (2011). The survey and future evolution of green computing. In: *2011 IEEE/ACM International Conference on Green Computing and Communications* (Vol. 1, pp. 230–233). IEEE.
38. Mankel, R., (1997). A concurrent track evolution algorithm for pattern recognition in the HERA-B main tracking system. *Nuclear Instruments and Methods in Physics Research Section A: Accelerators, Spectrometers, Detectors and Associated Equipment, 395*(2), 169–184.
39. Megino, F. B., De, K., Klimentov, A., Maeno, T., Nilsson, P., Oleynik, D., & ATLAS Collaboration. (2017). PanDA for ATLAS distributed computing in the next decade. In: *Journal of Physics: Conference Series* (Vol. 898, No. 5, p. 052002). IOP Publishing.
40. Mehlführer, C., Wrulich, M., Ikuno, J. C., Bosanska, D., & Rupp, M., (2009). Simulating the long term evolution physical layer. In: *2009 17th European Signal Processing Conference* (Vol. 1, pp. 1471–1478). IEEE.
41. Migdalas, A., Toraldo, G., & Kumar, V., (2003). Nonlinear optimization and parallel computing. *Parallel Computing, 29*(4), 375–391.
42. Mikkilineni, R., & Sarathy, V., (2009). Cloud computing and the lessons from the past. In: *2009 18th IEEE International Workshops on Enabling Technologies: Infrastructures for Collaborative Enterprises* (Vol. 1, pp. 57–62). IEEE.
43. Mikkilineni, R., Comparini, A., & Morana, G., (2012). The Turing o-machine and the DIME network architecture: Injecting the architectural resiliency into distributed computing. In: *Turing-100* (Vol. 1, pp. 239–251).

44. Milenkovic, M., Robinson, S. H., Knauerhase, R. C., Barkai, D., Garg, S., Tewari, A., & Bowman, M., (2003). Toward internet distributed computing. *Computer, 36*(5), 38–46.
45. Mühlenbein, H., Gorges-Schleuter, M., & Krämer, O., (1987). New solutions to the mapping problem of parallel systems: The evolution approach. *Parallel Computing, 4*(3), 269–279.
46. Mühlenbein, H., Gorges-Schleuter, M., & Krämer, O., (1988). Evolution algorithms in combinatorial optimization. *Parallel Computing, 7*(1), 65–85.
47. Mühlenbein, H., Schomisch, M., & Born, J., (1991). The parallel genetic algorithm as function optimizer. *Parallel Computing, 17*(6, 7), 619–632.
48. Omicini, A., & Viroli, M., (2011). Coordination models and languages: From parallel computing to self-organization. *The Knowledge Engineering Review, 26*(1), 53–59.
49. Padhy, R. P., & Patra, M. R., (2012). Evolution of cloud computing and enabling technologies. *International Journal of Cloud Computing and Services Science, 1*(4), 182.
50. Pelayo, F. L., Valverde, J. C., Pelayo, M. L., & Cuartero, F., (2010). Discrete dynamical systems for encoding concurrent computing systems. In: *9th IEEE International Conference on Cognitive Informatics (ICCI'10)* (Vol. 1, pp. 252–256). IEEE.
51. Poikolainen, I., & Neri, F., (2013). Differential evolution with concurrent fitness based local search. In: *2013 IEEE Congress on Evolutionary Computation* (Vol. 1, pp. 384–391). IEEE.
52. Rattner, J., (1985). Concurrent processing: A new direction in scientific computing. In: *Managing Requirements Knowledge, International Workshop* (Vol. 1, pp. 157). IEEE Computer Society.
53. Rimal, B. P., Choi, E., & Lumb, I., (2009). A taxonomy and survey of cloud computing systems. In: *2009 Fifth International Joint Conference on INC, IMS and IDC* (Vol. 1, pp. 44–51). IEEE.
54. Satyanarayanan, M., (1993). Mobile computing. *Computer, 26*(9), 81, 82.
55. Satyanarayanan, M., (1996). Fundamental challenges in mobile computing. In: *Proceedings of the Fifteenth Annual ACM Symposium on Principles of Distributed Computing* (Vol. 1, pp. 1–7).

56. Schmidt, H. A., Strimmer, K., Vingron, M., & Von, H. A., (2002). Tree-puzzle: Maximum likelihood phylogenetic analysis using quartets and parallel computing. *Bioinformatics, 18*(3), 502–504.
57. Schmitt, C., Rey-Coyrehourcq, S., Reuillon, R., & Pumain, D., (2015). Half a billion simulations: Evolutionary algorithms and distributed computing for calibrating the simpoplocal geographical model. *Environment and Planning B: Planning and Design, 42*(2), 300–315.
58. Shih, T. K., (2001). Mobile agent evolution computing. *Information Sciences, 137*(1–4), 53–73.
59. Skjellum, A., Smith, S. G., Doss, N. E., Leung, A. P., & Morari, M., (1994). The design and evolution of zipcode. *Parallel Computing, 20*(4), 565–596.
60. Stahl, D. O., & Whinston, A. B., (1994). A general economic equilibrium model of distributed computing. In: *New Directions in Computational Economics* (Vol. 1, pp. 175–189). Springer, Dordrecht.
61. Suciu, G., Halunga, S., Apostu, A., Vulpe, A., & Todoran, G., (2013). Cloud computing as evolution of distributed computing-a case study for SlapOS distributed cloud computing platform. *Informatica Economica, 17*(4), 109.
62. Sunderam, V. S., & Geist, G. A., (1999). Heterogeneous parallel and distributed computing. *Parallel Computing, 25*(13, 14), 1699–1721.
63. Sunderam, V. S., Geist, G. A., Dongarra, J., & Manchek, R., (1994). The PVM concurrent computing system: Evolution, experiences, and trends. *Parallel Computing, 20*(4), 531–545.
64. Tagawa, K., & Ishimizu, T., (2010). Concurrent differential evolution based on MapReduce. *International Journal of Computers, 4*(4), 161–168.
65. Tan, K. C., Tay, A., & Cai, J., (2003). Design and implementation of a distributed evolutionary computing software. *IEEE Transactions on Systems, Man, and Cybernetics, Part C (Applications and Reviews), 33*(3), 325–338.
66. Tasoulis, D. K., Pavlidis, N. G., Plagianakos, V. P., & Vrahatis, M. N., (2004). Parallel differential evolution. In: *Proceedings of the 2004 congress on evolutionary computation (IEEE Cat. No. 04TH8753)* (Vol. 2, pp. 2023–2029). IEEE.
67. Taylor, R. P., Berghaus, F., Brasolin, F., Cordeiro, C. J. D., Desmarais, R., Field, L., & ATLAS Collaboration, (2015). The evolution of cloud

computing in ATLAS. In: *Journal of Physics: Conference Series* (Vol. 664, No. 2, p. 022038). IOP Publishing.

68. Umemoto, K., Endo, A., & Machado, M., (2004). From sashimi to Zen-in: The evolution of concurrent engineering at fuji xerox. *Journal of Knowledge Management, 1*, 2–8.
69. Venkat, R. A., Oussalem, Z., & Bhattacharya, A. K., (2021). Training convolutional neural networks with differential evolution using concurrent task apportioning on hybrid CPU-GPU architectures. In: *2021 IEEE Congress on Evolutionary Computation (CEC)* (Vol. 1, pp. 2567–2576). IEEE.
70. Venkatasubramanian, N., (1999). CompOSE| Qa QoS-enabled customizable middleware framework for distributed computing. In: *Proceedings. 19th IEEE International Conference on Distributed Computing Systems; Workshops on Electronic Commerce and Web-Based Applications. Middleware* (Vol. 1, pp. 134–139). IEEE.
71. Vondrous, A., Selzer, M., Hötzer, J., & Nestler, B., (2014). Parallel computing for phase-field models. *The International Journal of High Performance Computing Applications, 28*(1), 61–72.
72. Vrugt, J. A., Nualláin, B. O., Robinson, B. A., Bouten, W., Dekker, S. C., & Sloot, P. M., (2006). Application of parallel computing to stochastic parameter estimation in environmental models. *Computers & Geosciences, 32*(8), 1139–1155.
73. Westerlund, M., & Kratzke, N., (2018). Towards distributed clouds: A review about the evolution of centralized cloud computing, distributed ledger technologies, and a foresight on unifying opportunities and security implications. In: *2018 International Conference on High Performance Computing & Simulation (HPCS)* (Vol. 1, pp. 655–663). IEEE.
74. Xia, M., Li, T., Zhang, Y., & De Silva, C. W., (2016). Closed-loop design evolution of engineering system using condition monitoring through internet of things and cloud computing. *Computer Networks, 101*, 5–18.
75. Ying, M., (2012). *Topology in Process Calculus: Approximate Correctness and Infinite Evolution of Concurrent Programs* (Vol. 1, pp. 2–6). Springer Science & Business Media.
76. Yokote, Y., & Tokoro, M., (1987). Experience and evolution of concurrent Smalltalk. In: *Conference Proceedings on Object-Oriented*

Programming Systems, Languages and Applications (Vol. 1, pp. 406–415).

77. Yong, L., Lishan, K., & Evans, D. J., (1995). The annealing evolution algorithm as function optimizer. *Parallel Computing, 21*(3), 389–400.
78. Zhu, W., (2009). A study of parallel evolution strategy: Pattern search on a GPU computing platform. In: *Proceedings of the First ACM/SIGEVO Summit on Genetic and Evolutionary Computation* (Vol. 1, pp. 765–772).

CHAPTER 3

CONCURRENT COMPUTING

CONTENTS

3.1. INTRODUCTION

Concurrent computing is natural and easier. The human brain behaves as a multiprocessor computer, which performs many tasks simultaneously. However, despite the fact we are good at processing parallel information, it is difficult to be aware of the activities we perform concurrently. When we try to raise awareness, we end up distracting ourselves and reducing the quality of what we are doing. Only after intense training can we, like a musician, conduct several activities simultaneously (Sunderam et al., 1994).

It is much easier to reason sequentially, doing only one thing at a time, than to understand situations where many things occur simultaneously. Furthermore, we are limited by our main communication mechanisms with others, spoken or written language, which are inherently sequential (Corradini, 1995). These convey information in parallel through the voice tone, facial, and body expressions, writing style, etc., but we are often unaware of it. Thus, while we are "parallel processors," and we leave in a world where multiple things happen at the same time, we usually reason by reduction to a sequential world. The same happens in computing. It is much easier to reason about a sequential program, than about one in which operations are executed concurrently (Figure 3.1) (Jou and Abraham, 1986).

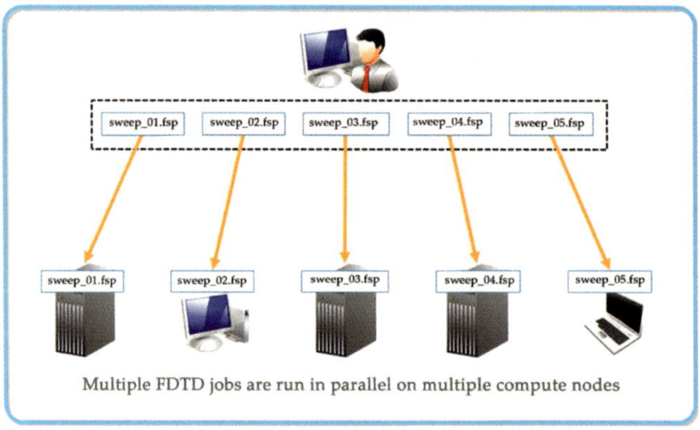

Figure 3.1. Concurrent computing.

Source: https://optics.ansys.com/hc/en-us/articles/360026162414-Concurrent-Parametric-Computing.

For more than 50 years, one of the most daunting challenges in information science and technology lies in mastering concurrency. Concurrency, once a specialized discipline for experts, is forcing itself onto the entire IT community because of two disruptive phenomena: the development of networking communications, and the end of the ability to increase processors speed at an exponential rate. Increases in performance can only come through concurrency, as in multicore architectures. Concurrency is also critical to achieve fault-tolerant, distributed available services, as in distributed data bases and cloud computing (Rajsbaum and Raynal, 2019). Yet, software support for these advances' lags, mired in concepts from the 1960s such as semaphores. The problem is compounded by the inherently non-deterministic nature of concurrent programs: even minor timing variations may generate completely different behavior. Sometimes tricky forms of concurrency faults can appear, such as data races (where even though two concurrent threads handle a shared data item in a way that is correct from each thread's perspective, a particular run-time interleaving produces inconsistent results) (Manohar and Martin, 1998; Panchal et al., 2013), and others, such as deadlocks (situation where some threads still have computations to execute, while none of them can proceed as each is waiting for another one to proceed), improper scheduling of threads or processes, priority inversions (where some processes do not proceed because the resources they need are unduly taken away by others) and various kinds of failures of the processes and communication mechanisms. The result is that it is very difficult to develop concurrent applications. Concurrency is also the means to achieve high performance computing, but in this chapter we are not concerned with such applications (Geist and Sunderam, 1992).

What does concurrent computing through sequential thinking mean? Instead of trying to reason directly about concurrent computations, the idea is to transform problems in the concurrent domain into simpler problems in the sequential domain, yielding benefits for specifying, implementing, and verifying concurrent programs. It is a two-sided strategy, together with a bridge connecting them (Ferrari and Sunderam, 1995):

- Sequential specifications for concurrent programs;
- Concurrent implementations;
- Consistency conditions relating concurrent implementations to sequential specifications.

Although a program is concurrent, the specification of the object (or service) that is implementing is usually through a sequential specification, stating the desired behavior only in executions where the processes execute one after the other, and often trough familiar paradigms from sequential computing (such as queues, stacks, and lists). In the previous example, when we state the rule that if the balance drops below a certain threshold, L, some high interest will be charged, we are thinking of an account which is always in an atomic state, i.e., a state where the balance is well defined (Agha et al., 1993; Wujek et al., 1996). This makes it easy to understand the object being implemented, as opposed to a truly concurrent specification which would be hard or unnatural. Thus, instead of trying to modify the well-understood notion of say, a queue, we stay with the sequential specification, and move to another level of the system the meaning of a concurrent implementation of a queue (Shukur et al., 2020).

3.2. ELEMENTS OF THE APPROACH

3.2.1. A Historical Perspective

The history of concurrency is long and there is an enormous amount of development, on many perspectives. As far as concurrency in distributed computing is concerned, the interested reader will find many results (both in shared-memory and message passing) in textbooks such as (Rattner, 1985) to cite a few. Many more books exist on other approaches, such as systems, formal methods, and security. The focus here is the history about the theoretical ideas of sequential reasoning to master concurrency (Farhat and Wilson, 1987).

In early 1960's Dijkstra published his seminal work on multi-programming (Shukur et al., 2020) which gave rise to concurrent programming. At the end of the 1970's distributed computing had become a possibility. Soon after, the Internet was born and it was being used to send email and do file transfer (Yvonnet et al., 2009; Acharjee and Zabaras, 2006). Additionally, it was hoped that the Internet could be used to run applications distributed over many machines, but there was little understanding of how that could be accomplished. It was motivated by a lack of understanding basic theory, that theoretical distributed computing was born, the first ACM conference on *Principles of Distributed Computing* in 1982, and the first impossibility results such as the first version the famous FLP impossibility result in 1983, and the highlighting of lower bounds (e.g., (Farhat and Wilson, 1987)) (Rattner, 1985).

Already in 1967 there were debates about multi-processor computing and Amdahl's Law predicting speedup when using multiple processors. In the late 1970s a move occurred from multiprocessors with shared memory to multicomputers with distributed memory communicating by sending messages. In the 1990s, the importance of shared memory with multicores returns, as it meets the barriers of energy expenditure, and the limits of making processors increasingly faster (Fox, 1987; Usui et al., 2010), emphasizing that the exponential growth prophesied by Moore's Law refers to packaging more and more components on the same chip, that is more and more parallel computing. And a new era of distributed systems is entered, which is motivated by new distributed services such as ledger-based applications, which must be open, scalable, and tolerate arbitrarily malicious faults. Nevertheless, while both parallel computing and distributed computing involve concurrency, it is important to stress the fact they address different computing worlds (Farhat et al., 1987).

3.2.1.1. At the Origin

Mutual-exclusion to share physical resources. Concurrent computing began in 1961 with what was called multiprogramming in the Atlas computer, where concurrency was simulated – as we do when telling stories where things happen concurrently – interlacing the execution of sequential programs. Concurrency was born in order to make efficient use of a sequential computer, which can execute only one instruction at a time, giving users the illusion that their programs are all running simultaneously, through the operating system. A collection of early foundational articles on concurrent programming appears in (Farhat et al., 1987) (Milner, 1993).

As soon as the programs being run concurrently began to interact with each other, it was realized how difficult it is to think concurrently. By the end of the 1960s there was already talk of a crisis: programming was done without any conceptual foundation and lots of programs were riddled with subtle race-related errors causing erratic behaviors. In 1965, Dijkstra discovered that the *mutual exclusion* of parts of code is a fundamental concept of programming, and opened the way for the first books of principles on concurrent programming which appeared at the beginning of the 1970s (Chung et al., 2021).

Interestingly, a very early impossibility result, showing a lower bound on the size of memory needed for two-process mutual exclusion (Chung et al., 2021), was one of the motivations for the systematic study of distributed computing by pioneers of the field (see for example Ref. (Amon et al., 1995)) (Li et al., 2018).

3.2.1.2. Fault-Tolerance

Some of the earliest sources of the approach followed in this chapter come from fault tolerance concerns, dating back to the 1960's, and in fact going back to von Neumann, as described by Avizienis^ in (Breitkopf, 2014), where he describes the sources of fault-tolerant techniques and some techniques used up to day (Xia and Breitkopf, 2014; Amon et al., 1995): "Considerable effort has been continuously directed toward practical use of massive triple modular redundancy (TMR) in which logic signals are handled in three identical channels and faults are masked by vote-taking elements distributed throughout the system." The idea begins to emerge, that "The fact that a task is executed by several processors is invisible to the application software" (Ogata et al., 2001).

The primary backup approach to fault-tolerance is an elementary early way of producing the illusion of a single computer, originating in 1976 or earlier (Ogata et al., 2001). A survey of about 30 years on replication collects many techniques, some from the very practical perspective (Ogata et al., 2001). It is about replicating data or a service that will behave in a manner indistinguishable from the behavior of some non-replicated reference system running on a single non-faulty node. More modern techniques such as *virtual synchrony, viewstamped replication, quorums*, etc., have their origins in Lamport's seminal 1978 paper (Olson, 2008) proposing the fundamental *state machine replication* paradigm (McDowell and Olson, 2008).

3.2.1.3. Resource Sharing

In 1965, Robert Taylor at ARPA started to fund large programs in advanced research in computing at major universities and corporate research centers throughout the United States, especially motivated by resource sharing. Among the computer projects that ARPA supported was time-sharing, in which many users could work at terminals to share a single large computer The thing that drove Bob Taylor at ARPA was resource sharing (Farhat and Roux, 1991).

Time-sharing was developed in the late 1950's out of the realization that a single expensive computer could be efficiently utilized if a multitasking, multiprogramming operating system allowed multiple users simultaneous interactive access. The DEC's PDP-10 is the machine that made time-sharing common, and this and other features made it a common fixture in many university computing facilities and research labs during the 1970s. Time-sharing emergence as the prominent model of computing in the

1970's represented a major technological shift in the history of computing (Rodrigues et al., 2018).

3.2.2. Shared Objects

In concurrent computing, a problem is solved through several threads (processes) that execute a set of tasks. In general, and except in so called "embarrassingly parallel" programs, i.e., programs that solve problems that can easily and regularly be decomposed into independent parts, the tasks usually need to synchronize their activities by accessing shared constructs, i.e., these tasks depend on each other (Athas and Seitz, 1988; Rocha et al., 2021). These typically serialize the threads and reduce parallelism. According to Amdahl's law (Rocha et al., 2021), the cost of accessing these constructs significantly impacts the overall performance of concurrent computations. Devising, implementing, and making good usage of such synchronization elements usually lead to intricate schemes that are very fragile and sometimes error prone (Figure 3.2) (Seitz, 1984).

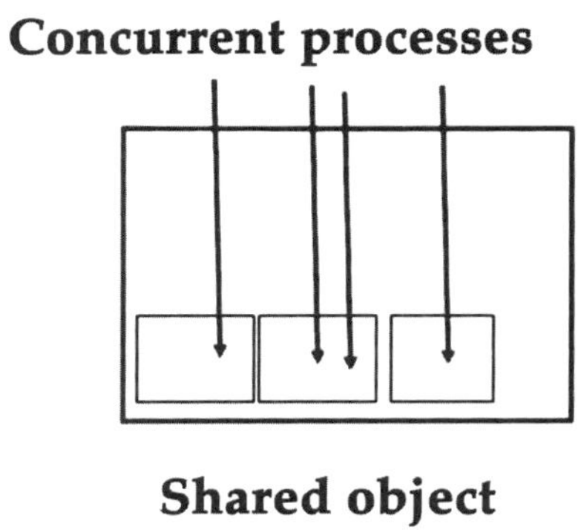

Figure 3.2. Concurrent computing and shared objects.

Source: https://www.slideshare.net/anniyappa/shared-objects-and-synchronization-28062313.

Every multicore architecture provides synchronization constructs in hardware. Usually, these constructs are "low-level" and making good usage of them is not trivial. Also, the synchronization constructs that are provided in hardware differ from architecture to architecture, making concurrent programs hard to port (Saether et al., 2007). Even if these constructs look the same, their exact semantics on different machines may also be different, and some subtle details can have important consequences on the performance or the correctness of the concurrent program. Clearly, coming up with a high-level library of synchronization abstractions that could be used across multicore architectures is crucial to the success of the multicore revolution. Such a library could only be implemented in software for it is simply not realistic to require multicore manufacturers to agree on the same high-level library to offer to their programmers (Pham and Dimov, 1998).

We assume a small set of low-level synchronization primitives provided in hardware, and we use these to implement higher level synchronization abstractions. As pointed out, these abstractions are supposed to be used by programmers of various skills to build application pieces that could themselves be used within a higher-level application framework (Lu et al., 2006).

The quest for synchronization abstractions, i.e., the topic of this book, can be viewed as a continuation of one of the most important quests in computing: programming abstractions. Indeed, the History of computing is largely about devising abstractions that encapsulate the specifities of underlying hardware and help programmers focus on higher level aspects of software applications (Sahu and Grosse, 1994). A file, a stack, a record, a list, a queue, and a set, are well-known examples of abstractions that have proved to be valuable in traditional sequential and centralized computing. Their definitions and effective implementations have enabled programming to become a high-level activity and made it possible to reason about algorithms without specific mention of hardware primitives (Rao et al., 1994).

The way an abstraction (object) is implemented is usually hidden to its users who can only rely on its operations and their specification to design and produce upper layer software, i.e., software using that object. The only visible part of an object is the set of values in can return when its operations are invoked. Such a modular approach is key to implementing provably correct software that can be reused by programmers in different applications (Naga and Agrawal, 2008).

The objects considered have a precise sequential specification, called also its sequential type, which specifies how the object behaves when accessed sequentially by the processes. That is, if executed in a sequential context (without concurrency), their behavior is known. This behavior might be deterministic in the sense that the final state and response is uniquely defined given every operation, input parameters and initial state. But this behavior could also be non-deterministic, in the sense that given an initial state of the object, and operation and an input parameter, there can be several possibilities for a new state and response (Synn and Fulton, 1995).

3.2.3. Reducibility of Algorithms

The notions of high and primitive being of course relative as we will see. It is also important to notice that the term implement is to be considered in an abstract manner; we will describe the concurrent algorithms in pseudo-code. There will not be any C or Java code in this book. A concrete execution of these algorithms would need to go through a translation into some programming language (Godefroid and Wolper, 1994).

An object to be implemented is typically called high-level, in comparison with the objects used in the implementation, considered at a lower-level. It is common to talk about emulations of the high-level object using the low-level ones. Unless explicitly stated otherwise, we will by default mean wait-free implementation when we write implementation, and atomic (linearizable) object when we write object (Shavit and Lotan, 2000).

It is often assumed that the underlying system provides some form of registers as base objects. These provide the abstraction of read-write storage elements. Message-passing systems can also, under certain conditions, emulate such registers. Sometimes the base registers that are supported are atomic but sometimes not. As we will see in this book, there are algorithms that implement atomic registers out of non-atomic base registers that might be provided in hardware (Yagawa and Shioya, 1993).

Some multiprocessor machines also provide objects that are more powerful than registers like test & set objects or compare & swap objects. These are more powerful in the sense that the writer process does not systematically overwrite the state of the object, but specifies the conditions under which this can be done. Roughly speaking, such conditional update enables more powerful synchronization schemes than with a simple register object. We will capture the notion of "more powerful" more precisely later in the book (Shih and Mitchell, 1992).

3.2.4. Correctness

Linearizability is a metric of the correctness of a shared object implementation. It addresses the question of what values can be returned by an object that is shared by concurrent processes. If an object returns a response, linearizability says whether this response is correct or not (Thomas, 1979).

The notion of correctness, as captured by linearizability, is defined with respect to how the object is expected to react when accessed sequentially: this is called the sequential specification of the object. In this sense, the notion of correctness of an object, as captured by linearizability, is relative to how the object is supposed to behave in a sequential world (Sun et al., 2018).

It is important to notice that linearizability does not say when an object is expected to return a response. As we will see later, the complementary to linearizability is the wait-freedom property, another correctness metric that captures the fact that an object operation should eventually return a response (if certain conditions are met) (Choi et al., 1992).

To illustrate the notion of linearizability, and the actual relation to a sequential specification, consider a FIFO (first-in-first-out) queue. This is an object of the type queue that contains an ordered set of elements and exhibits the following two operations to manipulate this set (Figure 3.3) (Farhat and Wilson, 1987):

- ***Enq*(*a*):** Insert element *a* at the end of the queue; and
- ***Deq*():** Return the first element inserted in the queue that was not already removed; Then remove this element from the queue; if the queue is empty, return the default element ⊥.

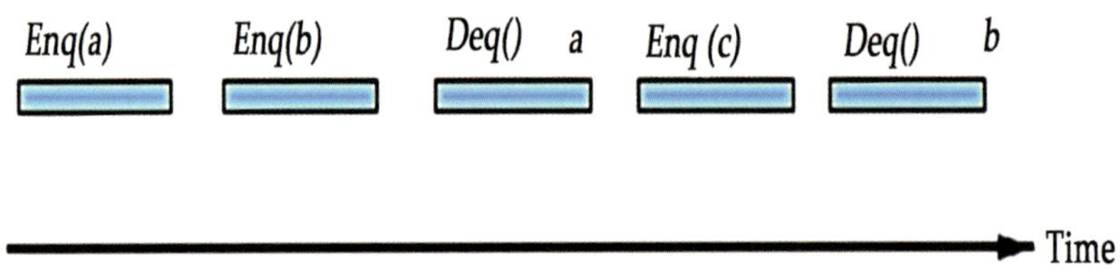

Figure 3.3. Sequential execution of a queue.

Source: https://www.amazon.com/Algorithms-for-concurrent-systems/dp/2889152839.

Figure 3.3 conveys a sequential execution of a system made up of a single process accessing the queue (here the time line goes from left to right). There is only a single object and a single process so we omit their identifiers here. The process first enqueues element *a*, then element *b*, and finally element *c*. According to the expected semantics of a queue (FIFO), and as depicted by the figure, the first dequeue invocation returns element *a* whereas the second returns element *b* (Rao et al., 1993).

Figure 3.4 depicts a concurrent execution of a system made up of two processes sharing the same queue: p_1 and p_2. Process p_2, acting as a producer, enqueues elements *a, b, c, d,* and then *e*. On the other hand, process p_1, acting as a consumer, seeks to de dequeue two elements. In Figure 3.4, the execution of *Enq*(*a*), *Enq*(*b*) and *Enq*(*c*) by p_2 overlaps with the first *Deq*() of p_1 whereas the execution of *Enq*(*c*), *Enq*(*d*) and *Enq*(*e*) by p_2 overlaps with the second *Deq*() of p_1. The questions raised in the figure are what elements can be dequeued by p_1. The role of linearizability is precisely to address such questions (Vernerey and Kabiri, 2012).

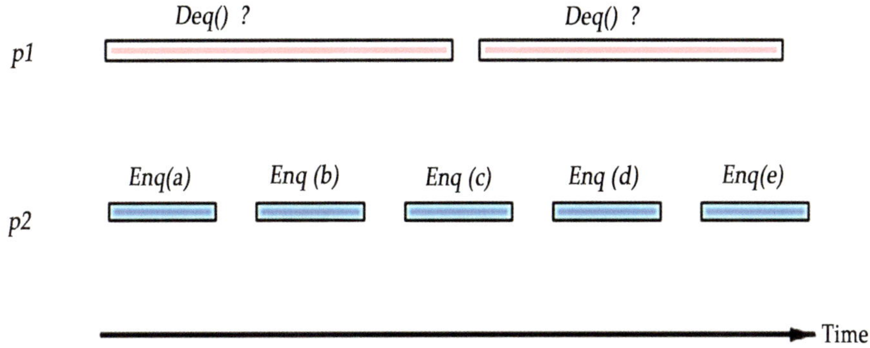

Figure 3.4. Concurrent execution of a queue.

Source: https://www.amazon.com/Algorithms-for-concurrent-systems/dp/2889152839.

Linearizability does so by relying on how the queue is supposed to behave if accessed sequentially. In other words, what should happen in Figure 3.4 depends on what happens in Figure 3.3. Intuitively, linearizability says that, when accessed concurrently, an object should return the same values that it could have returned in some sequential execution. Before defining linearizability however, and the very concept of "value that could have been returned in some sequential execution," we first define more precisely some important underlying elements, namely processes and objects, and then the very notion of a sequential specification (Farhat and Crivelli, 1989).

3.3. LINEARIZABILITY IN CONCURRENT COMPUTING

Essentially, linearizability says that a history is correct if the response returned to all operation invocations could have been obtained by a sequential execution, i.e., according to the sequential specifications of the objects. More specifically, we say that a history is linearizable if each operation appears as if it has been executed instantaneously at some indivisible point between its invocation event and its response event. This point is called the linearization point of the operation. We define below more precisely linearizability as well as some of its main characteristics (Saraswat et al., 1991).

3.3.1. Complete Histories

For pedagogical reasons, it is easier to first define linearizability for complete histories *H*, i.e., histories without pending operations, and then extend this definition to incomplete histories (Filipović et al., 2010).

- **Definition 1:** A complete history *H* is linearizable if there is a history *L* such that:

H and L are equivalent;

L is sequential;

L is legal; and

$\rightarrow H \subseteq \rightarrow L$.

Hence, a history H is linearizable if there exist a permutation of H, L, which satisfies the following requirements. First, L has to be indistinguishable from H to any process: this is the meaning of equivalence. Second, L should not have any overlapping operations: it has to be sequential. Third, the restriction of L to every object involved in it should belong to the sequential specification of that object: it has to be legal. Finally, L has to respect the real-time occurrence order of the operations in H (O'Hearn et al., 2010).

In short, L represents a history that could have been obtained by executing all the operations of H, one after the other, while respecting the occurrence order of non-overlapping operations in H. Such a sequential history L is called a linearization of H or a sequential witness of H (Herlihy and Wing, 1990).

An algorithm implementing some shared object is said to be linearizable if all histories generated by the processes accessing the object are linearizable. Proving linearizability boils down to exhibiting, for every such history, a linearization of the history that respects the "real-time" occurrence order of the operations in the history, and that belongs to the sequential specification of the object. This consists in determining for every operation of the history, its linearization point in the corresponding sequential witness history. To respect the real time occurrence order, the linearization point associated with an operation has always to appear within the interval defined by the invocation event and the response event associated with that operation. It is also important to notice that a history *H*, may have multiple possible linearizations (Schellhorn et al., 2014).

- *Example with a queue*: Consider history *H* depicted on Figure 3.5. Whether *H* is linearizable or not depends on the values returned by the dequeue invocations of p_1, i.e., in events e_7 and e_{13}. For example, assuming that the queue is initially empty, two possible values are possible for e_7: *a* and *nil* (Castañeda et al., 2015).
- In the first case, depicted on Figure 3.4, the linearization of the first dequeue of p_1 would be before the first enqueue of p_2. We depict a possible linearization on Figure 3.6 (Afek et al., 2010).

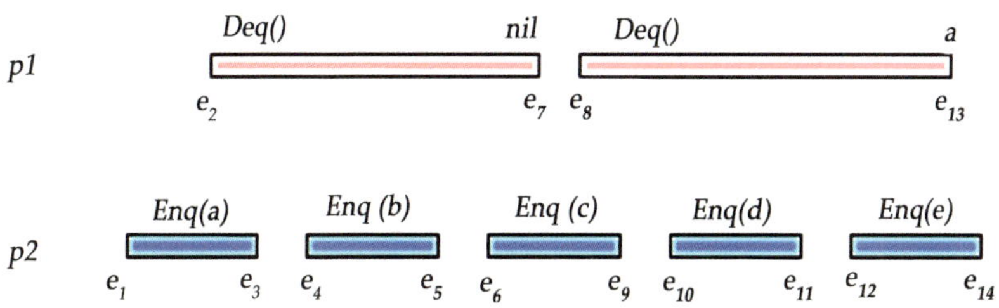

Figure 3.5. The first example of a linearizable history with queue.

Source: https://press.uchicago.edu/ucp/books/book/distributed/A/bo110495403.html.

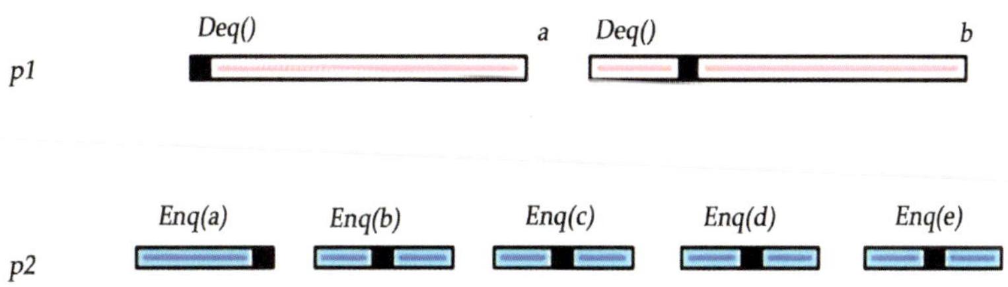

Figure 3.6. The first example of a linearization.

Source: https://press.uchicago.edu/ucp/books/book/distributed/A/bo110495403.html.

- In the second case, depicted on Figure 3.7, the linearization of the first dequeue of p_1 would be after the first enqueue of p_2. We depict a possible linearization on Figure 3.8 (Izraelevitz et al., 2016).

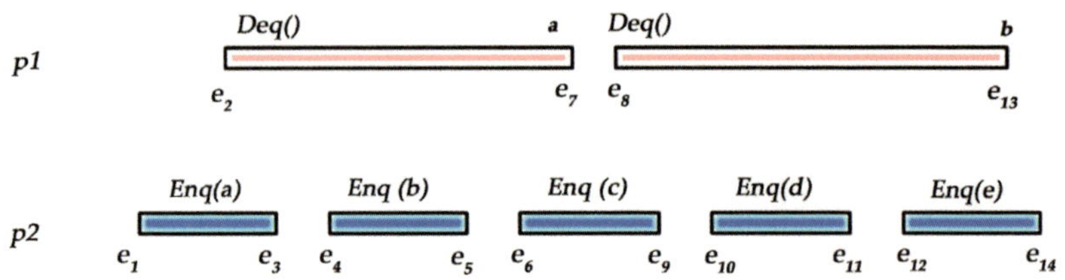

Figure 3.7. The second example of a linearizable history with a queue.

Source: https://press.uchicago.edu/ucp/books/book/distributed/A/bo110495403.html.

It is important to notice that, in order to ensure linearizability, the only possible values for e_7 are *a* and *nil*. If any other value was returned, the history of Figure 3.7 would not have been linearizable. For instance, if the value was *b*, i.e., if the first dequeue of p_1 returned *b*, then we could not have found any possible linearization of the history. Indeed, the dequeue should be linearizable after the enqueue of *b*, which is after the enqueue of *a*. To be legal, the linearization should have a dequeue of *a* before the dequeue of *b*—a contradiction (Haas et al., 2016).

- *Example with a register*: Figure 3.9 highlights a history of two processes accessing a shared register. The history contains events e_1, …, e_{12}. The history has no pending operations, and is consequently complete (Filipović et al., 2009).

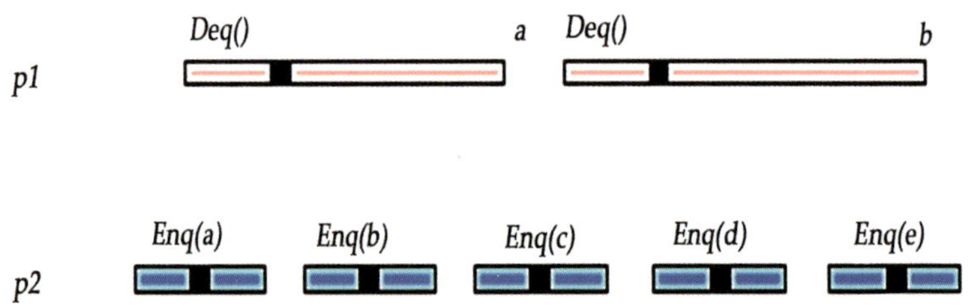

Figure 3.8. The second example of linearization.

Source: https://www.epflpress.org/produit/909/9782889152834/algorithms-for-concurrent-systems.

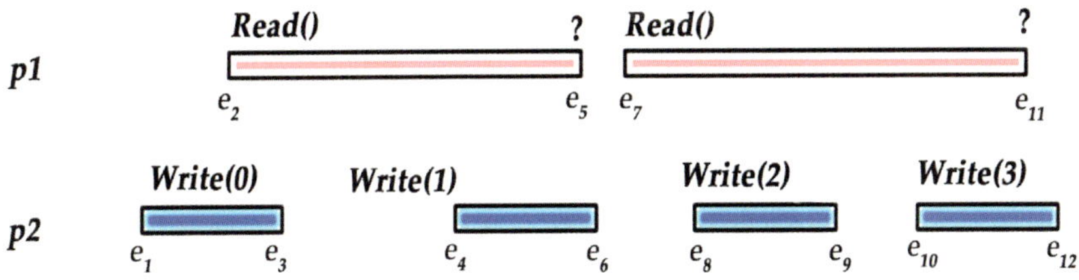

Figure 3.9. Example of a register history.

Source: *https://www.epflpress.org/produit/909/9782889152834/algorithms-for-concurrent-systems.*

Assuming that the register initially stores value 0, two possible returned values are possible for e_5 in order for the history to be linearizable: 0 and 1. In the first case, the linearization of the first read of p_1 would be right after the first write of p_2. In the second case, the linearization of the first read of p_1 would be right after the second write of p_2 (Zhang et al., 2013).

For the second read of p_1, the history is linearizable, regardless of whether the second read of p_1 returns values 1, 2 or 3 in event e_7. If this second read had returned a 0, the history would not be linearizable (Amit et al., 2007).

3.3.2. Incomplete Histories

So far, we considered only complete histories. These are histories with at least one process whose last operation is pending: the invocation event of this operation appears in the history while the corresponding response event does not. Extending linearizability to incomplete histories is important as it allows to state what responses are correct when processes crash. We cannot decide when processes crash and then cannot expect from a process to first terminate a pending operation before crashing (Travkin et al., 2013).

- **Definition 2:** A history H (whether it is complete or not) is linearizable if H can be completed in such a way that every invocation of a pending operation is either removed or completed with a response event, so that the resulting (complete) history H^0 is linearizable (Liu et al., 2009).

Basically, this definition transforms the problem of determining whether an incomplete history *H* is linearizable to the problem of determining whether a complete history H^0, obtained by completing *H*, is linearizable. H^0 is obtained by adding response events to certain pending operations of *H*, as if these operations have indeed been completed, or by removing invocation events from some of the pending operations of H. (All complete operations of H are preserved in H^0.) It is important to notice that the term "complete" is here a language abuse as we might "complete" a history by removing some of its pending invocations. It is also important to notice that, given an incomplete history H, we can complete it in several ways and derive several histories H^0 that satisfy the required conditions (Castañeda et al., 2018).

- *Example with a queue*: Figure 3.10 depicts an incomplete history H. We can complete H by adding to it the response b to the second dequeue of p_1 and a response to the last enqueue of p_2: we would obtain history H^0 which is linearizable. We could also have "completed" H by removing any of the pending operations, or both of them. In all cases, we would have obtained a complete history that is linearizable (Attiya and Welch, 1994).

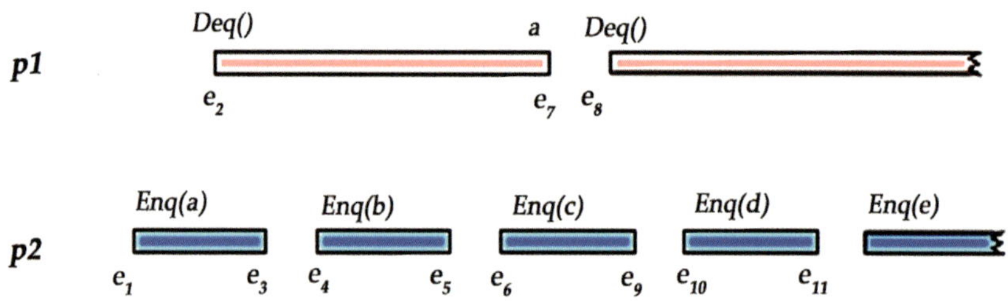

Figure 3.10. A linearizable incomplete history.

Source: https://perso.telecom-paristech.fr/kuznetso/WEP18/book-ln.pdf.

Figure 3.11 also depicts an incomplete history. However, no matter how we try to complete it, either by adding responses or removing invocations, there is no way to determine a linearization of the completed history.

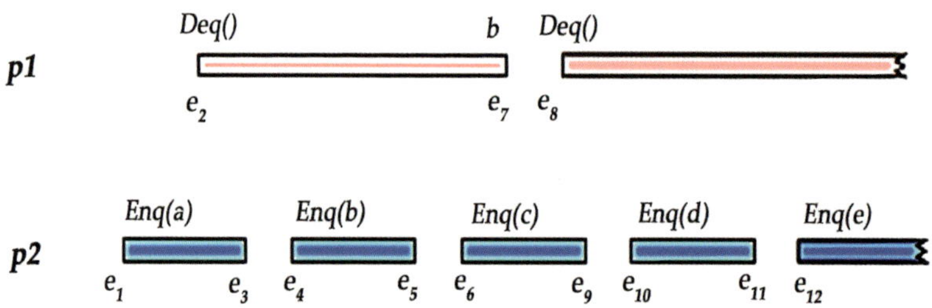

Figure 3.11. A non-linearizable incomplete history.

Source: https://perso.telecom-paristech.fr/kuznetso/WEP18/book-ln.pdf.

- *Example with a register:* Figure 3.12 depicts an incomplete history of a register. The only way to complete the history in order to make it linearizable is to complete the second write of p_2. This would enable the read of p_1 to be linearized right after it (Černý et al., 2010).

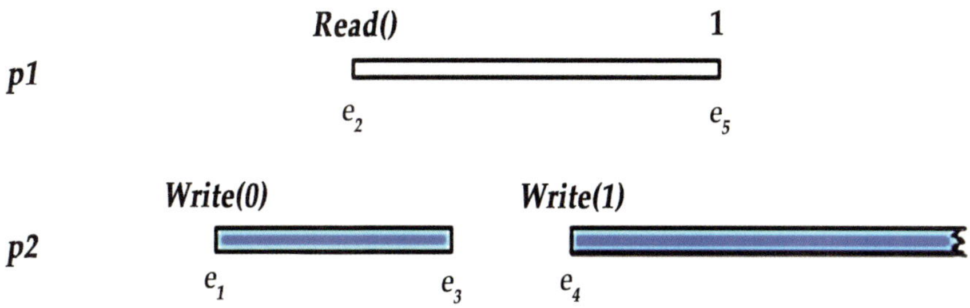

Figure 3.12. A linearizable incomplete history of a register.

Source: https://perso.telecom-paristech.fr/kuznetso/WEP18/book-ln.pdf.

3.4. COMPOSITION OF CONCURRENT SYSTEM

We discuss here a fundamental characteristic of linearizability as a *property*, i.e., as a set of histories. A property P is said to be *compositional* if whenever it holds for each of the objects of a set, it holds for the entire set. For each history H, we have $\forall X\ H|X \in P$ if and only if $H \in P$. A compositional property is also said to be *local*. Intuitively, compositionality enables to derive the correctness of a composed system from the correctness of the components. This property is crucial for modularity of programming: a correct (linearizable) composition can be obtained from correct (linearizable) components (Henrio et al., 2016).

Hence the transitive closure of $\rightarrow$ is irreflexive and anti-symmetric and, thus, has a linear extension: a total order on operations in H that respects $\rightarrow H$ and $\rightarrow X$, for all X. Consider the sequential history S induced by any such total order. Since, for all X, $S|X = S_X$ and S_X is legal, S is legal. Since $\rightarrow H \subseteq \rightarrow S$, S respects the real-time order of H. Finally, since each S_X is equivalent to a completion of $H|X$, S is equivalent to a completion of H, where each incomplete operation on an object X is completed in the way it is completed in S_X. Hence, S is a linearization of H. $2_{Theorem}$ (Klavins and Koditschek, 2000).

Linearizability stipulates correctness with respect to a sequential execution: an operation needs to appear to take effect instantaneously, respecting the sequential specification of the object. In this respect, linearizability is similar to sequential consistency, a classical correctness criterion for shared objects. There is however a fundamental difference between linearizability and sequential consistency, and this difference is crucial to making linearizability compositional, which is not the case for sequential consistently, as we explain below (Sun et al., 2019).

Sequential consistency is a relaxation of linearizability. It only requires that the real-time order is preserved if the operations are invoked by the same process, i.e., S is only supposed to respect the *process-order* relation.

More specifically, a history H is *sequentially consistent* if there is a "witness" history S such that:

- H and S are equivalent;
- S is sequential and legal.

Clearly, any linearizable history is also sequentially consistent. The contrary is not true. A major drawback of sequential consistency is that it is not compositional. To illustrate this, consider the counterexample described in Figure 3.13. The history H depicted in the picture involves two processes p_1 and p_2 accessing two shared registers R_1 and R_2. It is easy to see that the restriction H to each of the registers is sequentially consistent. Indeed, concerning register R_1, we can re-order the read of p_1 before the write of p_2 to obtain a sequential history that respects the semantics of a register (initialized to 0). This is possible because the resulting sequential history does not need to respect the real-time ordering of the operations in the original history. Note that the history restricted to R_1 is not linearizable. As for register R_2, we simply need to order the read of p_1 after the write of p_2 (Ananth et al., 2021).

Nevertheless, the system composed of the two registers R_1 and R_2 is not sequentially consistent. In every legal equivalent to H, the write on R_2 performed by p_2 should precede the read of R_2 performed by p_1: p_1 reads the value written by p_2. If we also want to respect the process-order relation of H on p_1 and p_2, we obtain the following sequential history: $p_2.Write_{R1}(1)$; $p_2.Write_{R2}(1)$; $p_1.Read_{R2}()$ 1; $p_1.Read_{R1}()$ 0. But the history is not legal: the value read by p_1 in R_1 is not the last written value, sequential history respecting the process-order relation of H must have (Wijs, 2013).

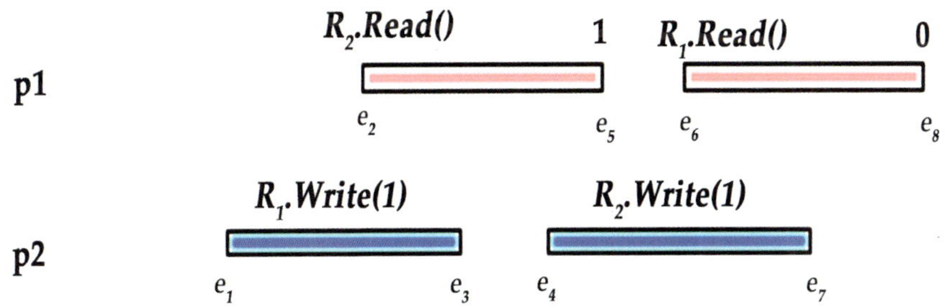

Figure 3.13. Sequential consistency is not compositional.

Source: https://perso.telecom-paristech.fr/kuznetso/WEP18/book-ln.pdf.

3.5. PROGRESS PROPERTY OF CONCURRENT COMPUTING

Linearizability (when applied to objects with finite non-determinism) is a safety property: it states what should not happen in an execution (Li et al., 2012).

Such a property is in fact easy to satisfy. Think of an implementation (of some shared object) that simply never returns any response. Since no operation would ever complete, the history would basically be empty and would be trivial to linearize: no response, no need for a linearization point. But this implementation would be useless. In fact, to prevent such implementations, we need a progress property stipulating those certain responses should appear in a history, at least eventually and under certain conditions. Ideally, we would like every invoked operation to eventually return a matching response. But this is impossible to guarantee if the process invoking the operation crashes, e.g., the process is paged out by the operating system which could decide not to schedule that process anymore (Henrio et al., 2016).

Nevertheless, one might require that a response is returned to a process that is scheduled by the operating system to execute enough steps of the algorithm implementing that operation (i.e., implementing the object exporting the operation). As we will see below, a step here is the access to a low-level object (used in the implementation) during the operation's execution (Cederman and Tsigas, 2012).

In the following, we introduce the notion of implementation history: this is a lower-level notion than the history notion which describes the interaction between the processes and the object being implemented (high-level history) The concept of low-level history will be used to introduce progress properties of shared object implementations (Canini et al., 2013).

3.6. IMPLEMENTATION

In order to reason about the very notion of implementation, we need to distinguish the very notions of high-level and low-level objects (Sun et al., 2019).

3.6.1. High-Level and Low-Level Objects

To distinguish the shared object to be implemented from the underlying objects used in the implementation, we typically talk about a high-level object and underlying low-level objects. (The latter are sometimes also called base objects and the operations they export are called primitives). That is, a process invokes operations on a high-level object and the implementation of these operations requires the process to invoke primitives of the underlying low-level (base) objects. When a process invokes such a primitive, we say that the process performs a step (Klavins and Koditschek, 2000).

The very notions of "high-level" and "low-level" are relative and depend on the actual implementation. An object might be considered high-level in a given implementation and low-level in another one. The object to be implemented is the high-level one and the objects used in the implementation are the low-level ones. The low-level objects might capture basic synchronization constructs provided in hardware and in this case the high-level ones are those we want to emulate in software (the notion of emulation is what we call implement). Such emulations are motivated by the desire to facilitate the programming of concurrent applications, i.e., to provide the programer with powerful synchronization abstractions encapsulated by high-level objects. Another motivation is to reuse programs initially devised with the high-level object in mind in a system that does not provide such an object in hardware. Indeed, multiprocessor machines do not all provide the same basic synchronization abstractions (Wijs, 2013).

Of course, an object O that is low-level in a given implementation A does not necessarily correspond to a hardware synchronization construct. Sometimes, this object O has itself been obtained from a software implementation B from some other lower objects. So O is indeed low-level in A and high-level in B. Also, sometimes the low-level objects are assumed to be linearizable, and sometimes not. In fact, we will even study implementations of objects that are not linearizable, as an intermediate way to build linearizable ones (Li et al., 2012).

3.6.2. Zooming into Histories

So far, we represent computations using histories, as sequences of events, each representing an invocation or a response on the object to be implemented, i.e., the high-level object (Canini et al., 2013).

3.6.2.1. Implementation History

In contrast, reasoning about progress properties requires to zoom into the invocations and responses of the lower-level objects of the implementations, on top of which the high-level object is built. Without such zooming we may not be able to distinguish a process that crashes right after invoking a high-level object operation and stops invoking low-level objects, from one that keeps executing the algorithm implementing that operation and invoking primitives on low-level objects. As we pointed out, we might want to require that the latter completes the operation by obtaining a matching response, but we cannot expect any such thing for the former. In this chapter, we will consider as an implementation history, the low-level history involving invocations and responses of low-level objects. This is a refinement of the higher-level history involving only the invocations and responses of the high-level object to be implemented (Dam, 1988).

Consider the example of a fetch-and-increment (counter) high-level-object implementation. As low-level objects, the implementation uses an infinite array *T*[, ..., ∞] of TAS (test-and-set) objects and a snapshot-memory object *my-inc*. The high-level history here is a sequence of invocation and response events of fetch-and-increment operations, while the low-level history (or implementation history) is a sequence of primitive events *read*(), *update*(), *snapshot*() and *test-and-set*().

3.6.2.2. The Two Faces of a Process

To better understand the very notion of a low-level history, it is important to distinguish the two roles of a process. On the one hand, a process has the role of a *client* that sequentially invokes operations on the high-level object and receives responses. On the other hand, the process also acts as a *server* implementing the operations. While doing so, the process invokes primitives on lower-level objects in order to obtain a response to the high-level invocation (Zhang and Luke, 2009).

It might be convenient to think of the two roles of a process as executed by different entities and written by two different programmers. As a client, a process invokes object operations but does not control the way the low-level primitives implementing these operations are executed. The programer writing this part does typically not know how an object operation is implemented. As a server, a process executes the implementation algorithm made up of invocations of low-level object primitives. This algorithm is typically written by a different programer who does not need to know what client applications will be using this object. Similarly, the client application does not need to know how the objects used are implemented, exccpt that they ensure linearizability and some progress property as discussed in further subsections (da Silva, 1970).

3.6.2.3. Scheduling and Asynchrony

The execution of a low-level object operation is called a *step*. The interleaving of steps in an implementation is specified by a *scheduler* (itself part of an operating system). This is outside of the control of processes and, in our context, it is convenient to think of a scheduler as an *adversary*. This is because, when devising an algorithm to implement some high-level object, one has to cope with worst-case strategies the scheduler may choose to defeat the algorithm. This is then viewed as an adversarial behavior (Shen et al., 2012).

A process is said to be correct in a low-level history if it executes an infinite number of steps, i.e., when the scheduler allocates infinitely many steps of that process. This "infinity" notion models the fact that the process executes as many steps as needed by the implementation until all responses are returned. Otherwise, if the process only takes finitely many steps, it is said to be faulty. In this book, we only assume that faulty processes crash, i.e., permanently stop performing steps, otherwise they never deviate from the algorithm assigned to them. In other words, they are not malicious (we also say they are not Byzantine) (Chandy and Taylor, 1989).

REFERENCES

1. Acharjee, S., & Zabaras, N., (2006). A concurrent model reduction approach on spatial and random domains for the solution of stochastic PDEs. *International Journal for Numerical Methods in Engineering, 66*(12), 1934–1954.
2. Afek, Y., Korland, G., & Yanovsky, E., (2010). Quasi-linearizability: Relaxed consistency for improved concurrency. In: *International Conference on Principles of Distributed Systems* (Vol. 1, pp. 395–410). Springer, Berlin, Heidelberg.
3. Agha, G., Frolund, S., Kim, W., Panwar, R., Patterson, A., & Sturman, D., (1993). Abstraction and modularity mechanisms for concurrent computing. *IEEE Parallel & Distributed Technology: Systems & Applications, 1*(2), 3–14.
4. Amit, D., Rinetzky, N., Reps, T., Sagiv, M., & Yahav, E., (2007). Comparison under abstraction for verifying linearizability. In: *International Conference on Computer Aided Verification* (Vol. 1, pp. 477–490). Springer, Berlin, Heidelberg.
5. Amon, C. H., Nigen, J. S., Siewiorek, D. P., Smailagic, A., & Stivoric, J., (1995). Concurrent design and analysis of the navigator wearable computer system: The thermal perspective. *IEEE Transactions on Components, Packaging, and Manufacturing Technology: Part A, 18*(3), 567–577.
6. Ananth, P., Chung, K. M., & Placa, R. L. L., (2021). On the concurrent composition of quantum zero-knowledge. In: *Annual International Cryptology Conference* (Vol. 1, pp. 346–374). Springer, Cham.
7. Athas, W. C., & Seitz, C. L., (1988). Multicomputers: Message-passing concurrent computers. *Computer, 21*(8), 9–24.
8. Attiya, H., & Welch, J. L., (1994). Sequential consistency versus linearizability. *ACM Transactions on Computer Systems (TOCS), 12*(2), 91–122.
9. Canini, M., Kuznetsov, P., Levin, D., & Schmid, S., (2013). Software transactional networking: Concurrent and consistent policy composition. In: *Proceedings of the Second ACM SIGCOMM Workshop on Hot Topics in Software Defined Networking* (Vol. 1, pp. 1–6).
10. Castañeda, A., Rajsbaum, S., & Raynal, M., (2015). Specifying concurrent problems: Beyond linearizability and up to tasks. In:

International Symposium on Distributed Computing (Vol. 1, pp. 420–435). Springer, Berlin, Heidelberg.

11. Castañeda, A., Rajsbaum, S., & Raynal, M., (2018). Unifying concurrent objects and distributed tasks: Interval-linearizability. *Journal of the ACM (JACM), 65*(6), 1–42.
12. Cederman, D., & Tsigas, P., (2012). Supporting lock-free composition of concurrent data objects: Moving data between containers. *IEEE Transactions on Computers, 62*(9), 1866–1878.
13. Černý, P., Radhakrishna, A., Zufferey, D., Chaudhuri, S., & Alur, R., (2010). Model checking of linearizability of concurrent list implementations. In: *International Conference on Computer Aided Verification* (Vol. 1, pp. 465–479). Springer, Berlin, Heidelberg.
14. Chandy, K. M., & Taylor, S., (1989). The composition of concurrent programs. In: *Proceedings of the 1989 ACM/IEEE Conference on Supercomputing* (Vol. 1, pp. 557–561).
15. Choi, J., Dongarra, J. J., Pozo, R., & Walker, D. W., (1992). ScaLAPACK: A scalable linear algebra library for distributed memory concurrent computers. In: *The Fourth Symposium on the Frontiers of Massively Parallel Computation* (Vol. 1, pp. 120–121). IEEE Computer Society.
16. Chung, P. W., Tamma, K. K., & Namburu, R. R., (2001). Asymptotic expansion homogenization for heterogeneous media: Computational issues and applications. *Composites Part A: Applied Science and Manufacturing, 32*(9), 1291–1301.
17. Corradini, A., (1995). Concurrent computing: From petri nets to graph grammars. *Electronic Notes in Theoretical Computer Science, 2*, 56–70.
18. da Silva, A. F., (1970). Visual composition of concurrent applications. *WIT Transactions on Information and Communication Technologies, 5*, 2–9.
19. Dam, M., (1988). Relevance logic and concurrent composition. In: *LICS* (Vol. 88, pp. 178–185).
20. Farhat, C., & Crivelli, L., (1989). A general approach to nonlinear FE computations on shared-memory multiprocessors. *Computer Methods in Applied Mechanics and Engineering, 72*(2), 153–171.
21. Farhat, C., & Roux, F. X., (1991). A method of finite element tearing and interconnecting and its parallel solution algorithm. *International Journal for Numerical Methods in Engineering, 32*(6), 1205–1227.

22. Farhat, C., & Wilson, E., (1987). A new finite element concurrent computer program architecture. *International Journal for Numerical Methods in Engineering, 24*(9), 1771–1792.
23. Farhat, C., & Wilson, E., (1987). Concurrent iterative solution of large finite element systems. *Communications in Applied Numerical Methods, 3*(4), 319–326.
24. Farhat, C., Wilson, E., & Powell, G., (1987). Solution of finite element systems on concurrent processing computers. *Engineering with Computers, 2*(3), 157–165.
25. Ferrari, A. J., & Sunderam, V. S., (1995). TPVM: Distributed concurrent computing with lightweight processes. In: *Proceedings of the Fourth IEEE International Symposium on High Performance Distributed Computing* (Vol. 1, pp. 211–218). IEEE.
26. Filipović, I., O'Hearn, P., Rinetzky, N., & Yang, H., (2009). Abstraction for concurrent objects. In: *European Symposium on Programming* (Vol. 1, pp. 252–266). Springer, Berlin, Heidelberg.
27. Filipović, I., O'Hearn, P., Rinetzky, N., & Yang, H., (2010). Abstraction for concurrent objects. *Theoretical Computer Science, 411*(51, 52), 4379–4398.
28. Fox, G. C., (1987). The Caltech concurrent computation program. In: *Hypercube Multiprocessors, 1987: Proceedings of the Second Conference on Hypercube Multiprocessors, Knoxville, Tennessee* (Vol. 29, p. 353). SIAM.
29. Geist, G. A., & Sunderam, V. S., (1992). Network-based concurrent computing on the PVM system. *Concurrency: Practice and Experience, 4*(4), 293–311.
30. Godefroid, P., & Wolper, P., (1994). A partial approach to model checking. *Information and Computation, 110*(2), 305–326.
31. Haas, A., Henzinger, T. A., Holzer, A., Kirsch, C., Lippautz, M., Payer, H., & Veith, H., (2016). Local linearizability for concurrent container-type data structures. *Leibniz International Proceedings in Informatics, 1*, 59.
32. Henrio, L., Madelaine, E., & Zhang, M., (2016). A theory for the composition of concurrent processes. In: *International Conference on Formal Techniques for Distributed Objects, Components, and Systems* (Vol. 1, pp. 175–194). Springer, Cham.

33. Herlihy, M. P., & Wing, J. M., (1990). Linearizability: A correctness condition for concurrent objects. *ACM Transactions on Programming Languages and Systems (TOPLAS), 12*(3), 463–492.
34. Izraelevitz, J., Mendes, H., & Scott, M. L., (2016). Linearizability of persistent memory objects under a full-system-crash failure model. In: *International Symposium on Distributed Computing* (Vol. 1, pp. 313–327). Springer, Berlin, Heidelberg.
35. Jou, J. Y., & Abraham, J. A., (1986). Fault-tolerant matrix arithmetic and signal processing on highly concurrent computing structures. *Proceedings of the IEEE, 74*(5), 732–741.
36. Klavins, E., & Koditschek, D. E., (2000). A formalism for the composition of concurrent robot behaviors. In: *Proceedings 2000 ICRA; Millennium Conference; IEEE International Conference on Robotics and Automation; Symposia Proceedings (Cat. No. 00CH37065)* (Vol. 4, pp. 3395–3402). IEEE.
37. Li, H., Luo, Z., Gao, L., & Qin, Q., (2018). Topology optimization for concurrent design of structures with multi-patch microstructures by level sets. *Computer Methods in Applied Mechanics and Engineering, 331*, 536–561.
38. Li, H., Zhu, Q., Yang, X., & Xu, L., (2012). Geo-information processing service composition for concurrent tasks: A QoS-aware game theory approach. *Computers & Geosciences, 47*, 46–59.
39. Liu, Y., Chen, W., Liu, Y. A., & Sun, J., (2009). Model checking linearizability via refinement. In: *International Symposium on Formal Methods* (Vol. 1, pp. 321–337). Springer, Berlin, Heidelberg.
40. Lu, G., Tadmor, E. B., & Kaxiras, E., (2006). From electrons to finite elements: A concurrent multiscale approach for metals. *Physical Review B, 73*(2), 024108.
41. Manohar, R., & Martin, A. J., (1998). Slack elasticity in concurrent computing. In: *International Conference on Mathematics of Program Construction* (Vol. 1, pp. 272–285). Springer, Berlin, Heidelberg.
42. McDowell, D. L., & Olson, G. B., (2008). Concurrent design of hierarchical materials and structures. In: *Scientific Modeling and Simulations* (Vol. 1, pp. 207–240). Springer, Dordrecht.
43. Milner, R., (1993). Elements of interaction: Turing award lecture. *Communications of the ACM, 36*(1), 78–89.

44. Naga, B. K. G., & Agrawal, V. P., (2008). Concurrent design of a computer network for x-abilities using MADM approach. *Concurrent Engineering, 16*(3), 187–200.

45. O'Hearn, P. W., Rinetzky, N., Vechev, M. T., Yahav, E., & Yorsh, G., (2010). Verifying linearizability with hindsight. In: *Proceedings of the 29th ACM SIGACT-SIGOPS Symposium on Principles of Distributed Computing* (Vol. 1, pp. 85–94).

46. Ogata, S., Lidorikis, E., Shimojo, F., Nakano, A., Vashishta, P., & Kalia, R. K., (2001). Hybrid finite-element/molecular-dynamics/electronic-density-functional approach to materials simulations on parallel computers. *Computer Physics Communications, 138*(2), 143–154.

47. Panchal, J. H., Kalidindi, S. R., & McDowell, D. L., (2013). Key computational modeling issues in integrated computational materials engineering. *Computer-Aided Design, 45*(1), 4–25.

48. Pham, D. T., & Dimov, S. S., (1998). An approach to concurrent engineering. *Proceedings of the Institution of Mechanical Engineers, Part B: Journal of Engineering Manufacture, 212*(1), 13–27.

49. Rajsbaum, S., & Raynal, M., (2019). Mastering concurrent computing through sequential thinking. *Communications of the ACM, 63*(1), 78–87.

50. Rao, A. R. M., Loganathan, K., & Raman, N. V., (1993). Confess—An architecture-independent concurrent finite element package. *Computers & Structures, 47*(2), 343–348.

51. Rao, A. R. M., Loganathan, K., & Raman, N. V., (1994). Multi-frontal-based approach for concurrent finite element analysis. *Computers & Structures, 52*(4), 841–846.

52. Rattner, J., (1985). Concurrent processing: A new direction in scientific computing. In: *Managing Requirements Knowledge, International Workshop* (Vol. 1, pp. 157). IEEE Computer Society.

53. Rocha, I. B. C. M., Kerfriden, P., & Van, D. M. F. P., (2021). On-the-fly construction of surrogate constitutive models for concurrent multiscale mechanical analysis through probabilistic machine learning. *Journal of Computational Physics: X, 9*, 100083–100090.

54. Rodrigues, E. A., Manzoli, O. L., Bitencourt, Jr. L. A., Bittencourt, T. N., & Sánchez, M., (2018). An adaptive concurrent multiscale model for concrete based on coupling finite elements. *Computer Methods in Applied Mechanics and Engineering, 328*, 26–46.

55. Saether, E., Yamakov, V., & Glaessgen, E., (2007). A statistical approach for the concurrent coupling of molecular dynamics and finite element methods. In: *48th AIAA/ASME/ASCE/AHS/ASC Structures, Structural Dynamics, and Materials Conference* (Vol. 1, p. 2169).
56. Sahu, K., & Grosse, I. R., (1994). Concurrent iterative design and the integration of finite element analysis results. *Engineering with Computers, 10*(4), 245–257.
57. Saraswat, V. A., Rinard, M., & Panangaden, P., (1991). The semantic foundations of concurrent constraint programming. In: *Proceedings of the 18th ACM SIGPLAN-SIGACT Symposium on Principles of Programming Languages* (Vol. 1, pp. 333–352).
58. Schellhorn, G., Derrick, J., & Wehrheim, H., (2014). A sound and complete proof technique for linearizability of concurrent data structures. *ACM Transactions on Computational Logic (TOCL), 15*(4), 1–37.
59. Seitz, C. L., (1984). Concurrent VLSI architectures. *IEEE Transactions on Computers, 33*(12), 1247–1265.
60. Shavit, N., & Lotan, I., (2000). Skiplist-based concurrent priority queues. In: *Proceedings 14th International Parallel and Distributed Processing Symposium; IPDPS 2000* (Vol. 1, pp. 263–268). IEEE.
61. Shen, Y., Yang, X., Wang, Y., & Ye, Z., (2012). Optimizing QoS-aware services composition for concurrent processes in dynamic resource-constrained environments. In: *2012 IEEE 19th International Conference on Web Services* (Vol. 1, pp. 250–258). IEEE.
62. Shih, F. C., & Mitchell, O. R., (1992). A mathematical morphology approach to Euclidean distance transformation. *IEEE Transactions on Image Processing, 1*(2), 197–204.
63. Shukur, H., Zeebaree, S. R., Ahmed, A. J., Zebari, R. R., Ahmed, O., Tahir, B. S. A., & Sadeeq, M. A., (2020). A state of art survey for concurrent computation and clustering of parallel computing for distributed systems. *Journal of Applied Science and Technology Trends, 1*(4), 148–154.
64. Sun, M., Zhou, Z., Wang, J., Du, C., & Gaaloul, W., (2019). Energy-efficient IoT service composition for concurrent timed applications. *Future Generation Computer Systems, 100*, 1017–1030.

65. Sun, W., Fish, J., & Dhia, H. B., (2018). A variant of the s-version of the finite element method for concurrent multiscale coupling. *International Journal for Multiscale Computational Engineering, 16*(2), 2–9.
66. Sunderam, V. S., Geist, G. A., Dongarra, J., & Manchek, R., (1994). The PVM concurrent computing system: Evolution, experiences, and trends. *Parallel Computing, 20*(4), 531–545.
67. Synn, S. Y., & Fulton, R. E., (1995). Practical strategy for concurrent substructure analysis. *Computers & Structures, 54*(5), 939–944.
68. Thomas, R. H., (1979). A majority consensus approach to concurrency control for multiple copy databases. *ACM Transactions on Database Systems (TODS), 4*(2), 180–209.
69. Travkin, O., Mütze, A., & Wehrheim, H., (2013). SPIN as a linearizability checker under weak memory models. In: *Haifa Verification Conference* (Vol. 1, pp. 311–326). Springer, Cham.
70. Usui, T., Behrends, R., Evans, J., & Smaragdakis, Y., (2010). Adaptive locks: Combining transactions and locks for efficient concurrency. *Journal of Parallel and Distributed Computing, 70*(10), 1009–1023.
71. Vernerey, F. J., & Kabiri, M., (2012). An adaptive concurrent multiscale method for microstructured elastic solids. *Computer Methods in Applied Mechanics and Engineering, 241*, 52–64.
72. Wijs, A., (2013). Define, verify, refine: Correct composition and transformation of concurrent system semantics. In: *International Workshop on Formal Aspects of Component Software* (Vol. 1, pp. 348–368). Springer, Cham.
73. Wujek, B. A., Renaud, J. E., Batill, S. M., & Brockman, J. B., (1996). Concurrent subspace optimization using design variable sharing in a distributed computing environment. *Concurrent Engineering, 4*(4), 361–377.
74. Xia, L., & Breitkopf, P., (2014). Concurrent topology optimization design of material and structure within FE2 nonlinear multiscale analysis framework. *Computer Methods in Applied Mechanics and Engineering, 278,* 524–542.
75. Yagawa, G., & Shioya, R., (1993). Parallel finite elements on a massively parallel computer with domain decomposition. *Computing Systems in Engineering, 4*(4–6), 495–503.

76. Yvonnet, J., Gonzalez, D., & He, Q. C., (2009). Numerically explicit potentials for the homogenization of nonlinear elastic heterogeneous materials. *Computer Methods in Applied Mechanics and Engineering, 198*(33–36), 2723–2737.
77. Zhang, L., Chattopadhyay, A., & Wang, C., (2013). Round-Up: Runtime checking quasi linearizability of concurrent data structures. In: *2013 28th IEEE/ACM International Conference on Automated Software Engineering (ASE)* (Vol. 1, pp. 4–14). IEEE.
78. Zhang, Y., & Luke, E., (2009). Concurrent composition using loci. *Computing in Science & Engineering, 11*(3), 27–35.

CHAPTER 4

PARALLEL COMPUTING

CONTENTS

4.1. INTRODUCTION

Parallel computing varies widely amongst enterprise computer processors and they might or might not be physically coupled. For instance, one might have the same storage while others have just localized (private) storage, and their activity might well be coordinated by a similar clock or operate independently. Moreover, in the actual creation and operation of a computer, architectural intricacies and hardware specifications of the parts often manifest themselves. Lastly, there are technical distinctions, which present themselves in various clock speeds, shared memory times, etc. Whether the characteristics of parallel computing should be incorporated in the development as well as assessment of parallel algorithms, as well as which qualities might well be disregarded (Asanovic et al., 2009)?

To address the issue, comparable concepts to some of those identified for serial computing are used. Here, several computational respondents were designed. In brief, the purpose of all of such models was to separate the related from the unimportant qualities of the (sequential) computation (Padua, 2011).

In our particular instance, the random-access machine (RAM) design is especially appealing. Why? RAM summarizes the important characteristics of general-purpose sequenced computer systems, that are still widely utilized and which could be adopted as the fundamental concept for designing parallel computing as well as parallel computers. Figure 4.1 depicts the RAM's structure (Garland et al., 2008).

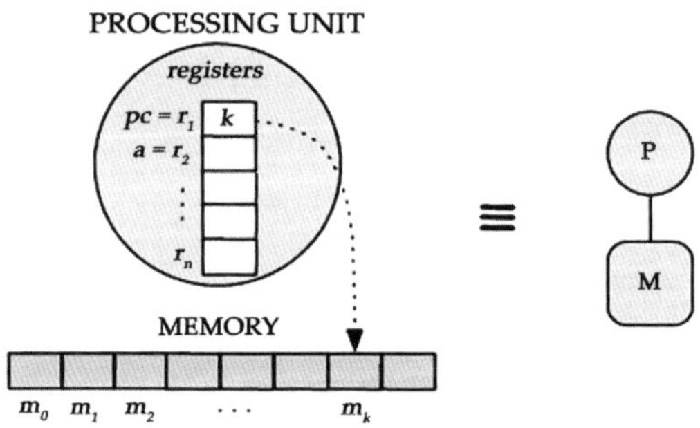

Figure 4.1. The RAM model of computation with a storage M (comprising program configuration information) as well as a processing unit P (working effectively on information).

Source: https://hpc.llnl.gov/documentation/tutorials/introduction-parallel-computing-tutorial.

Here is a concise explanation of RAM (Dongarra et al., 2003):

- The RAM is comprised of a processor and storage. Memory is a possibly unlimited series of places m0 and m1 of equivalent length. The index I is known as mi's address. Every position is instantly available by the processing system: a random I mi may be read from or written to in regular frequency. Records are a series of places r_1, …, r_n inside the processing unit. Records are readily available. Two of them play unique functions. The program counter pc ($=r_1$) carries the value of the memory area containing the next command to be performed. In every instruction execution, the accumulator a ($=r_2$) is used. Other registers are assigned functions as necessary. The software is a discrete series of commands (same as that in real computers).
- By starting the RAM, the preceding is performed: (a) a program is stored into consecutive main memory beginning with, like, m_0; (b) input data are put into vacant main memory, say, following the final program command.

It was discovered that answering this issue is far more difficult than in the situation of sequential processing. Why? As there are different methods of making parallel computers, there are also several methods to represent them. Choosing a singular model that is capable for everyone parallel computers is challenging (Sharma and Martin, 2009).

As a consequence, scholars have developed numerous models of parallel processing over the past few decades. Nonetheless, no consensus has emerged regarding the best option. The next section describes RAM-based systems (Barney, 2010).

4.2. MULTIPROCESSOR MODELS

A multiprocessor prototype is a parallel computation prototype that develops on the RAM computation prototype; multiprocessor designs. Each of the three models has some number p (2) of processing units however, the models are dissimilar in their entity, that is, it incorporates the RAM. How would it accomplish this (Waintraub et al., 2009)?

It did turn that out our description could be accomplished in three fundamental ways, consisting of 3 three distinct remembrances as well as three distinct ways for the computer processors to connect the memories. The designs are known as the parallel random-access memory (PRAM) (Li and Martinez, 2005):

- Local memory machine (LMM);
- Modular memory machine (MMM); and
- Parallel random-access memory (PRAM).

4.2.1. The Parallel Random-Access Machine (PRAM)

The PRAM model, or parallel random-access machine, has p processing units all of which are linked to a familiar unbridled memory space (Figure 4.2). By releasing a memory connection request to the memory space, so every processing element can connect such a position (word) in the memory space in a single step.

In many ways, the PRAM design of parallel computation is approximated. First, except for p, there is no boundary to the number of processing elements. The presumption that a processing element could indeed obtain every position in shared memory in a tiny step is also optimistic. Eventually, words in memory space are only presumed to be a similar size; or else, they could be of any fixed size (Ezhova and Sokolinskii, 2019).

It should be noted that there is no connectivity channel in this prototype for transmitting memory queries and information among both processing elements as well as the memory unit. (It will be much different in another two versions, the LMM and the MMM.)

The idea that every processing element could reach any memory address in a single movement, on the other hand, is unreasonable. To understand why, imagine that computer processors P_i and P_j concurrently concern guidelines I_i and I_j, both of which plan to connect (read from or write to) the similar memory location L (see Figure 4.3) (Drozdowski, 1996).

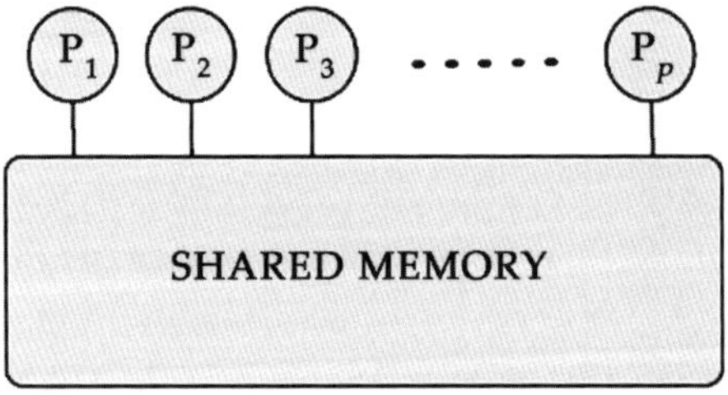

Figure 4.2. The parallel processing PRAM model: p processor units share an infinite amount of memory. So, every processing element could indeed connect every memory address in a single movement.

Source: https://hpc.llnl.gov/documentation/tutorials/introduction-parallel-computing-tutorial.

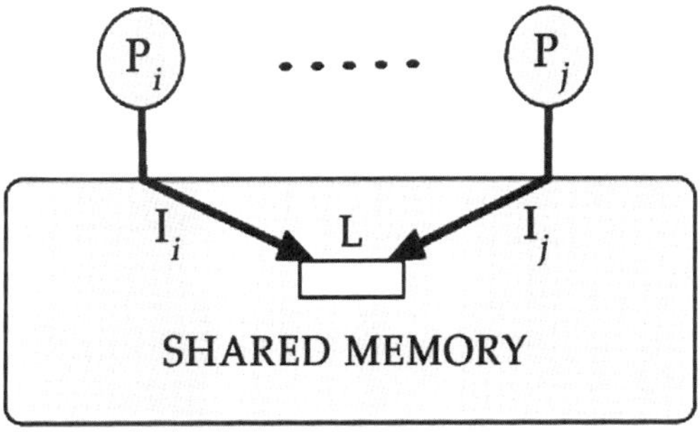

Figure 4.3. The dangers of continuity of service to a position. Two computer processors update the instruction set, each one must connect to a similar location L.

Source: https://hpc.llnl.gov/documentation/tutorials/introduction-parallel-computing-tutorial.

Although if true concurrent accessibility to L had been feasible, the components of L could have been unexpected. Consider the components of L after concurrently composing 3 and 5 into this one. As a result, it is fair to presume that real I_i and I_j get access to L are ultimately syndicated (sequentialized) by equipment such that I_i and I_j physiologically obtain L one after another (Orlov and Efimushkina, 2016).

As a result, the results of commands I_i and I_j are similarly unexpected (Figure 4.3). Why? When both P_i and P_j desire to take from L at a similar time, irrespective of serialization, the commands I_i and I_j will both access the identical data of L, therefore both computer processors will get the similar contents of L—as intended. When one of the processing elements would like to read from L while another wishes to create to L, the information captured by the receiving processing system will be determined by how much the reading command was serialized during the writing. Furthermore, when both P_i and P_j attempt to write into L at the same time, the eventual values of L would be determined by how I_i and I_j were serialized, — in other words, whichever of I_i and I_j was the final to commit to L (Veltman et al., 1990).

To summarize, concurrent reach to a similar place can result in unexpected information in both the requesting processing units and the requested location (Padua, 2011).

4.2.1.1. The Variants of PRAM

Due to the aforementioned problems, scientists devised many PRAM variants that vary in (Luo and Friedman, 1990):

- The types of concurrent attempts to access a similar location that are permitted; and
- the method for avoiding uncertainty whenever simultaneously attempting to access a similar location.

The variants are referred to as:

- EREW-PRAM (exclusive read exclusive write PRAM);
- Concurrent Read Exclusive Write PRAM (CREW-PRAM)
- Concurrent read concurrent write PRAM (CRCW-PRAM).

We will go over them in the most specific now (Wilkinson, 2006):

- **EREW-PRAM:** It is the most believable of the 3 PRAM model variants. Concurrent availability to a similar memory location is not supported by the EREW-PRAM design; if this is attempted, the model will end implementing its program. As a result, the obvious implication is that applications run on EREW-PRAM do not ever display the main indication that connects a similar memory location at a similar time; that is, all main memory has to be unique. As a result, algorithm developers are in charge of creating of that kind program.
- **CREW-PRAM:** The above model enables the possibility of mentions from a similar memory location but only allows for the exclusive writes to that memory area. The algorithm developer bears the brunt of the responsibility for creating these programs (Li, 1988).
- **CRCW-PRAM:** The PRAM model in this edition is the lowest accurate of the three. Concurrent readings from a similar memory region, concurrent puts to a similar memory location, as well as simultaneous reads and puts to a similar memory area, are all possible with the CRCW-PRAM paradigm. Therefore, distinct further constraints on concurrent transfers are enforced to prevent unforeseen results. As a result, the model CRCW-PRAM has the following variations (Ezhova and Sokolinskii, 2018):
- **CRCW-PRAM-Consistent:** Although many processing units might try to write to L at a similar time, it is expected that they

must all put a similar value to L. The algorithm writer must ensure this.

- **Arbitrary-CRCW-PRAM:** Computer processors may try writing to L at a similar time (though not always with a similar value), and only one was allowed to achieve. Because it is impossible to foresee which processing element would win, the programer should include it in machine learning (ML) (Culler et al., 1996).
- **Priority-CRCW-PRAM:** The computer processors have a priority ranking; for example, the processing element with the lower value has bigger importance. Although multiple processing modules may begin to write to L at the same time, it is presumed that only the one with top priority would then find success. Once again, the algorithm developer must anticipate and consider every potential outcome all through implementation (Andrews and Polychronopoulos, 1991).

4.2.1.2. The Relative Power of the Variants

The versions of PRAM are getting becoming less practical whenever the limits on continuity of service to a similar area are removed as we go from EREW-PRAM to CREW-PRAM and eventually to CRCW-PRAM. However, when the limits are removed, it is normal to predict that the variations will get more powerful (Dennis, 1994).

- **Proof Concept:** We first demonstrate that EREW-PRAM(p) can conduct CONSISTENT-CRCW-PRAM (simultaneous)'s writes to the same spot in O(log p) steps. As a result, EREW-PRAM may mimic CRCW-PRAM with a slowness factor of O. (log p). Next, we demonstrate that this slowness factor is constrained, i.e., that there is still a computing issue whereby the slowness component is genuinely constrained (log p). The challenge of determining the highest of n numbers is a case of this (Culler et al., 1993).
- **Relevance of the PRAM Model:** We have discussed why the PRAM model's presumption of an instantly installable, endless memory space is unachievable. Is this to say that the PRAM model is not relevant when it comes to using parallel computing execution? The answer is contingent on our expectations of the PRAM model or, more broadly, our understanding of the theory's role (Li and Martinez, 2006).

Whenever we try to create an algorithm for finding a solution to the problem on PRAM, we might not even finish up with a working algorithm that can be used to solve the problem. The layout, on the other hand, might very well indicate something, namely that it is parallelizable. In other words, the design may detect subproblems, some of which could be solved in parallel, at least in theory. It usually shows that such subproblems can be solved in parallel on the most liberal (and unrealistic) PRAM, the CRCW-PRAM, in this case (Kessler and Keller, 2007).

In conclusion, the following technique reflects the importance of PRAM (Wu et al., 1991):

- Create a program P for solving on the CRCW-PRAM(p) model, where p varies depending on the issue. Notice that designing P for CRCW-PRAM should be simpler than designing P for EREW-PRAM since CRCWPRAM does not have any concurrent constraints to consider.
- Execute P on EREW-PRAM(p), which is thought to be capable of simulating many concurrent requests to a similar place.
- Apply Theorem 2.1 to ensure that the parallel processing time of P on EREWPRAM(p) is no more than O(log p) times those from the less accurate CRCW-PRAM (p).

4.2.2. The Local-Memory Machine

Every processor unit in the LMM architecture does have its memory unit (Figure 4.4). The units are interconnected through a shared network. Every processor unit has immediate access to its memory location. Accessing non-local memory (i.e., the local storage of some other processor unit) requires submitting a memory demand across the network interface (Bhandarkar and Arabnia, 1995).

The presumption is that local activities, involving main memory, take a certain amount of time. In comparison, the time needed to reach non-local storage relies on the following factors (Sunderam, 1990):

- the capacity of the network interface;
- the sequence of concurrent non-local memory locations by other processing elements, since these requests may cause congestion on the network interface.

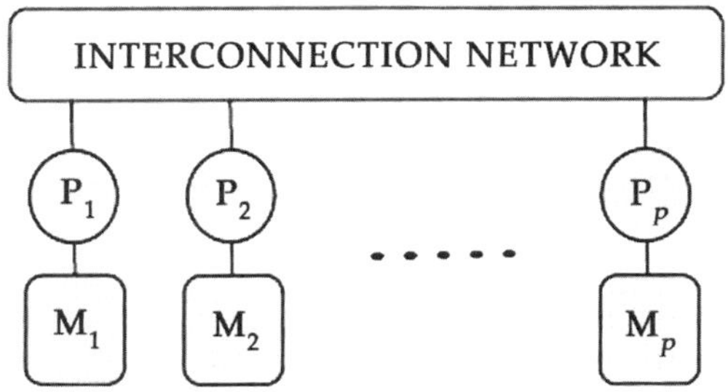

Figure 4.4. The LMM model of parallel computing contains p processing elements, within each memory location, as shown. Every processor unit has immediate links to its memory space and it may reach the local storage of other processing elements through a network interface.

Source: https://www.geeksforgeeks.org/introduction-to-parallel-computing/.

4.2.3. The Memory-Module Machine

The MMM architecture (Figure 4.5) is comprised of p processing elements and m main memory, all of which may be accessible either by processing system over a shared network interface. Processing units do not include local memories. A processing element may get entry to the storage drive by transmitting a memory demand over the network interface (Mukherjee and Bennett, 1990).

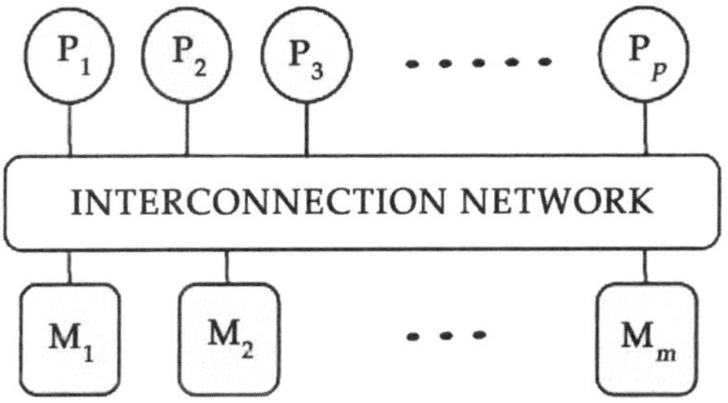

Figure 4.5. The MMM paradigm of parallel computing is shown with p processing elements as well as m main memory. Through the interconnectivity network, any processor may reach any memory card. There has been no functional unit storage.

Source: https://www.geeksforgeeks.org/introduction-to-parallel-computing/.

It is believed that the processing elements and primary storage are organized so that, in the absence of concurrent requests, the time required for any processing system to reach any memory card is approximately constant. Nevertheless, when concurrent connections occur, the accessing time is dependent on (Skillicorn, 1990):

- the connectivity network's capabilities; and
- the arrangement of memory accesses that occur simultaneously.

4.3. THE IMPACT OF COMMUNICATION

Either the LMM system or the MMM model employs open interconnect systems to transmit memory demands to non-local stores (see Figures 4.4 and 4.5). That part focuses on the importance of a network interface in the multiprocessing architecture and its effect on the computation time of parallel processing (Drozdowski, 1997).

4.3.1. Interconnection Networks

Because of the beginning of parallel computing, the primary characteristics of a parallel system have been the central processing unit (CPU) and interconnection network. This one is happening presently. Recent investigations have demonstrated that the implementation durations of the majority of actual parallel programs are based on time available instead of computation time. Consequently, when the quantity of collaborating processing elements or computers rises, the efficiency of network interfaces becomes more significant than that of the processing system. In particular, the interconnection network has a significant influence on the productivity and sustainability of a parallel computer for the majority of parallel uses in reality. In other terms, the maximum reliability of a network interface may eventually result in greater speedups, as it can reduce the total parallel completion time and expand the number of processing elements that can be utilized effectively (Vella, 1998).

The efficiency of a network interface is dependent on several variables. Forwarding, flow-control methods, and topology of the network are three of the most crucial aspects of a system. In this context, routing is the method of choosing a route for vehicles in a network interface; flow control is a method of handling the transmission rate among both pairs of nodes to avoid a quick message from dominating a slow recipient; as well as the topology of the network is the configuration of the different pieces, including interaction nodes as well as networks, of a network interface (Park et al., 2009).

For connectivity and flow-control techniques, there are already effective systems in place. In comparison, network topologies have not adapted as rapidly to shifting technology trends as route as well as flow-control algorithms. It is one of the reasons why several network topologies found shortly after the emergence of parallel processing still is frequently utilized today. Another argument is that users are allowed to choose the network configuration that best suits their projected needs. (As a result of contemporary standards, there is no such flexibility to choose or modify routing or flow-control algorithms.) As a result, it may be anticipated that changes in the topology of network interfaces would result in another improved performance. For instance, these enhancements should allow interconnecting systems to constantly and optimally adjust to the current implementation (Krämer and Mühlenbein, 1989).

4.3.2. Basic Properties of Interconnection Networks

We may categorize and describe interconnected networks using a variety of factors. The most attractive mathematical approach for specifying some of such characteristics is graph theory. Furthermore, an interconnected network may be represented by the graph G (N, C), where N represents a collection of interaction nodes as well as C represents a range of communication connections (or channels) connecting the communication branches. Based on such a graph-theoretic perspective on network interfaces, we may develop variables that capture both the structural and performance characteristics of network interfaces. Let us discuss two types of real estate (Sorel, 1994).

4.3.2.1. Topological Characteristics of Network Interconnections

By graph-theoretical concepts, the most essential topological aspects of interconnection networks are:

- Node degree;
- Pattern;
- Uniformity;
- Size;
- Route variety; and
- Extensibility scalability.

In the preceding, we explain every term and provide relevant commentary (Li and Zhang, 2018):

- The number of nodes is the number of connections via which an information node is linked to other transmission nodes. Note that just the message passing channels are included in the network density, although a communicating node also requires connections for the link to the processor element(s) as well as connections for service or support routes.
- A network is considered to be normal if all connection branches have a similar node degree, or if d > 0 so that every transmission node has a similar node degree.
- A network is symmetrical if all nodes have the "identical perspective" of the network, or when there is a particular behavior that translates each information node to other connected nodes. In a system with symmetric connections, the burden may be dispersed uniformly across all transmission nodes, hence eliminating congestion issues. Numerous actual installations of network interfaces are built on symmetrical normal graphs due to their advantageous topological qualities, which result in straightforward routes and equitable packet forwarding under homogeneous load.
- To proceed from a data packet to a destination network, a packet must cross a sequence of components, such as routers or switches, that make up the route (or path) between the sender and recipient. Hop count corresponds to the number of connection links passed by the package somewhere along the route. In an ideal scenario, network entities interact via the route with the smallest hop count, l, among all pathways connecting them. Because l might vary depending on the input and output ports, we also utilize the distance traveled, l_{avg}, which would be the mean of l for all potential node pairings. The dimension, l_{max}, which would be the sum of all of the lowest number of hops for all combinations of input and output ports, is an essential property of any topology (Benini et al., 2005).
- In a network interface, several pathways might occur between two nodes. In this scenario, there are several methods to link the nodes. A packet leaving the data signal will still have numerous pathways means of reaching the endpoint. Based on the present

state of the network, the packet might travel several paths (or indeed alternative extensions of a previously traversed path). A network with a high degree of route variety provides additional options for packets seeking their targets and/or avoiding barriers.

- Scalability is I the capacity of a system to manage an increasing volume of work, or (ii) the system's capacity to be expanded to meet this increase. Scalability is essential at all levels. For instance, the fundamental element must consistently link to other components. In addition, a similar building block should be used to construct interconnecting networks of various sizes, with only a little performance loss for the largest parallel computer. The influence of interconnection networks on the scalability of parallel computers based on the LMM or MMM multiprocessor paradigm is significant. Note that scalability is restricted if node degree is constant to comprehend this (Axelrod, 1986).

4.3.2.2. Interconnection System Performance Characteristics

The most key quality characteristics of network interfaces are:

- Channel capacity;
- Bandwidth for bisection;
- Latency.

We will now describe all of them and provide remarks if needed (Covington et al., 1991):

- Channel bandwidth, or simply bandwidth, is the quantity of information that can be transmitted across a route in given time duration. In most situations, the channel bandwidth could be sufficiently calculated by the simple communication design that promotes that the communication period t_{comm}, interacting, and communicating data sets thru the channel, is the total amount $t_s + t_d$ of the start-up period ts and the data transmission time t_d, where $t_d = mt_w$, the result of the word count creating up the information, m, and the transfer time one per term, t_w. The channel bandwidth is thus $1/t_w$.
- An interconnected network may be divided into two (nearly) displays. This may be accomplished in a variety of ways. The cut-bandwidth of a network interface is the total of the stream bandwidths among all routes linking the two parts. The

interconnection network's bisection bandwidth (BBW) is the littlest cut-bandwidth. The comparable cut is the connection network's worst-case cut. The bisection bandwidth per node (BBWN) is sometimes required; we describe it as BBW split by |N|, the count of branches in the system. Both BBW and BBWN are affected by network structure and channel bandwidths. Overall, increasing the interconnection network bandwidth may be as helpful as raising the CPU clock (Francis and Pannan, 1992).

- Latency is the amount of time it takes for a package to transit from the resource to the destination station. Some applications, particularly those using brief messages, are latency-sensitive in the view that their efficiency is highly dependent on delay. For these kinds of systems, software cost could become a significant element influencing delay. Finally, the latency is limited by the time it takes light to travel the actual distance between two terminals.

The data transmitted from a source address to a destination network is defined in light of the specific units (Dhillon and Modha, 2002):

- Package, the shortest quantity of data that technology can send;
- FLIT (flow control digit), is the quantity of data utilized in certain flow-control systems to assign storage space;
- The quantity of data that may be sent in a single period (PHIT).

These units are intimately connected to networking bandwidth as well as latency (Bader and Cong, 2005).

Although if described in a hypothetical greater geometry, each given network interface must finally be translated into an actual three-dimensional (3D) area. This implies that all of the chips, as well as the printed circuit that comprises the connectivity system, should be assigned physical locations (Guan et al., 2002).

However, that is no easy feat. The explanation for this is that mapping frequently needs to maximize specific, sometimes contradictory, criteria even while adhering to numerous constraints. Here have been a few such good examples (Tosic, 2004):

- One such limitation would be that the count of I/O pins on each chip or per circuit board is limited. A common optimization requirement is that wires be as little as feasible to avoid bit rate decreases. However, because of the larger dimensions of hardware devices and the actual restrictions of 3D space, the map might significantly extend some pathways, i.e., terminals that are near in higher-dimensional space might well be transferred to far places in 3D space.
- We might wish to put intensely communicating processor units as near to each other as possible, preferably on a similar chip. In this method, we might reduce the influence of information. However, building such ideal translations is an NP-hard optimization task.
- Another requirement could be that energy usage is kept to a minimum.

4.3.3. Classification of Interconnection Networks

Indirect and direct connections are two types of interconnection networks. The key characteristics of each kind are discussed in subsections (Peir, 1986).

4.3.3.1. Direct Networks

When every node in a system is linked directly to its neighbors, it also seems to be direct. What is the maximum number of neighbors a unit can still have? Every one of the n = |N| nodes in such a fully linked network is intrinsically linked to all of the other nodes, giving every node n 1 neighbors (see Figure 4.6) (Grindley et al., 2000).

Because a network contains n(n1) = (n2) direct interconnections, it could only be utilized to construct systems having n nodes. Whenever n is big, every node is connected to a set of many other networks, but data are routed via intermediary nodes to reach the other endpoints. The hypercube is a case of a direct interconnection network (Lin and Snyder, 1990).

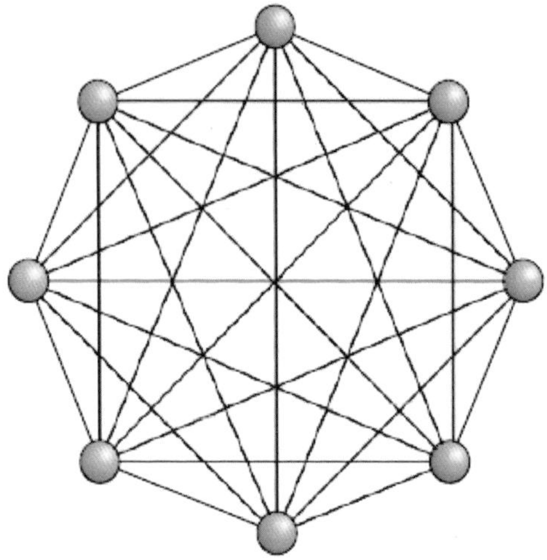

Figure 4.6. A completely linked network with n = 8 nodes.

Source: https://www.geeksforgeeks.org/introduction-to-parallel-computing/.

4.3.3.2. Indirect Networks

An indirect network links the unit using switches. It usually links the processing elements on one node of the network to the system memory on another. The completely linked cross switching is the easiest circuitry for linking processor processors to system memory (Figure 4.7). It has the benefit of being able to link processor modules and main memory in a variety of ways (Nishiura and Sakaguchi, 2011).

A cross point is found at the junction of two azimuth and elevation lines. A cross point is a tiny valve that may be opened () or closed (•) electrically based on both the vertical and lateral lines that need to be joined or not. Eight cross points are shut at a similar time in Figure 4.7, permitting linkages between both the pairings (P_1, M_1), (P_2, M_3), (P_3, M_5), (P_4, M_4), (P_5, M_2), (P_6, M_6), (P_7, M_8) and (P_8, M_7). There is a plethora of different possibilities (Kistler et al., 2006).

However, the fully linked crossbar is just too complicated to link a huge number of inner and outer ports. The quantity of cross-connections increases as pm, where p and m represent the number of processing elements and main memory, correspondingly. This translates to a million cross points for p = m = 1000, which is not possible. (However, a crossbar architecture is feasible for medium-sized systems, and tiny fully linked cross switching are utilized as fundamental building pieces inside bigger switches and routers) (Corcoran and Wainwright, 1994).

This is why indirect systems need several switches to link the units. The switching is normally linked to one another in phases, with a consistent communication sequence among them. Multistage connection systems are a kind of indirect network (Khan et al., 2008).

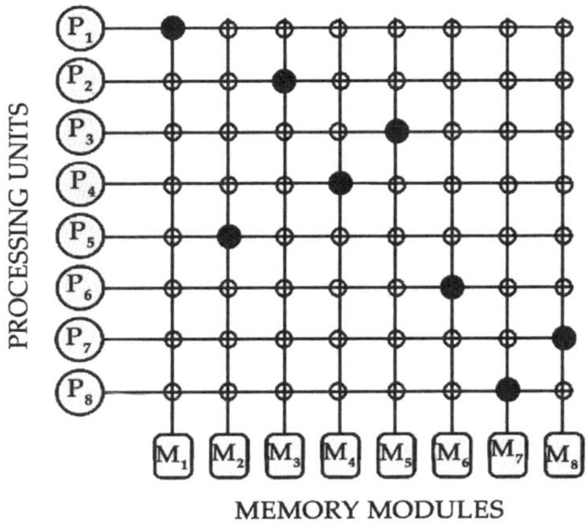

Figure 4.7. A completely linked crossbar switch linking eight nodes to eight nodes.

Source: https://www.amazon.com/Introduction-Parallel-Computing-Ananth-Grama/dp/0201648652.

Indirect networks are further divided into the following categories (Sokolinsky, 2021):

- *Irrespective* of the links previously developed throughout the network, a non-blocking network may link any idle origin to any inactive destination. This is because of the network architecture, which assures that many pathways remain between the sender and recipient.
- *A blocking* rearrangeable net may reorganize the links that have previously been formed throughout the network, allowing for the establishment of a new link. This kind of network may create any number of connections across outputs and inputs.
- *In a blocked network,* even though the input and output are both free, a link that has been created from across the system may

prevent the construction of a new link among them. A network of this kind cannot always link a source to an unspecified unconstrained endpoint.

Currently, the boundary between direct and indirect connections is blurry. Because each node in a direct system may be described as a gateway from its processing system linked to other circuits, any direct system could be described as an indirect system. The entire crossbar, as an electronic switch, is the core of the communications system both for direct and indirect connectivity networks (Rosing et al., 1992).

4.3.4. Topologies of Interconnection Networks

Numerous network topologies suitable for joining p processor modules and m memory devices are readily apparent (see Exercises). Nevertheless, not all computer networks are capable to transmit memory demands rapidly enough just to support parallel computations effectively (Guan et al., 2007). Furthermore, the network topology has a substantial effect on the effectiveness of the network interface and, therefore, parallel processing. The network topology may impose substantial challenges on the construction phase and expense of the network (Arge et al., 2008).

Scientists have created, researched, created, evaluated, and utilized different network topologies during the last several decades. The bus, the grid, the 3D-mesh, the torus, the hypercube, the multi-stage network, and the fat-tree are now described in detail (Mishra and Dehuri, 2011).

4.3.4.1. The Bus

It is the basic topology for a network (see Figure 4.8). It is compatible with both LMMs and memory-module machines (MMMs). All processor modules plus main memory are coupled to a bus line for either situation. One item of data may be put onto the bus in every phase. This may be a demand from a processing system to read and comprehend a memory value, or the reply from the processing system or memory component holding the value (Gropp and Lusk, 1995).

Whenever a processor unit in a memory-module computer wishes to retrieve a memory term, it must also determine whether the bus is occupied. If the bus is empty, the processing element places the requested word's address on the bus, sends the required control signals, and waits for the memory to place the required letter on the bus (Squillante and Lazowska, 1993). Nonetheless, if the bus is in use whenever a processing element desires to write or read memory, the processing element must delay till the bus is idle. Here, the shortcomings of the bus topology become evident. If there are just two or three processing units, congestion for the bus is tolerable; but, if there are 32 processing units, contention becomes insufferable since the majority of processing units will be waiting the majority of the time (Bracker et al., 1996).

To remedy this issue, a regional cache is added to each processor unit. The cache may be situated on the CPU board, adjacent to the CPU chip, within the CPU chip, or in a mix of all three locations. In practice, caching is not performed on a word-by-word basis, but rather based on 64-byte blocks. When a processing unit references a word, the word's whole block is retrieved into the processing unit's local cache (Poletti et al., 2007).

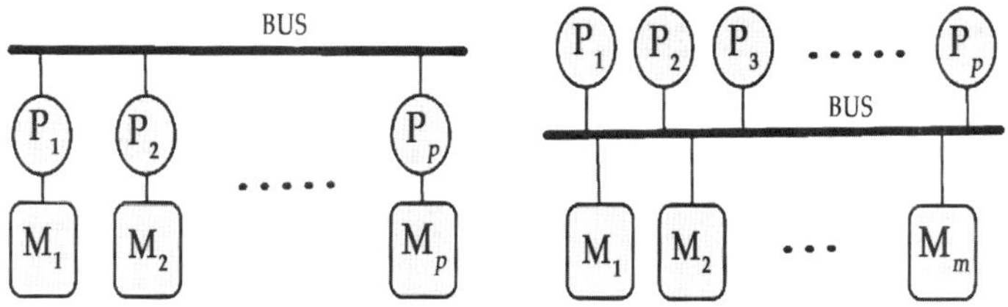

Figure 4.8. The bus is the basic network architecture.

Source: https://www.amazon.com/Introduction-Parallel-Computing-Ananth-Grama/dp/0201648652.

After then, a large number of reads may be supplied by the local cache. As a consequence, bus traffic would be reduced and the system would be able to accommodate additional processing machines (Li and Martinez, 2005).

The practical benefits of employing buses include their ease of construction and (ii) the relative simplicity of developing protocols that enable processing units to cache memory data regionally (since all processors and memory units can notice the traffic on the bus). Using a bus has the apparent drawback that the computer processors need alternate reaching the bus. This indicates that when additional processing units are added to a bus, the average time required to accomplish memory access increases proportionally (Gajski and Peir, 1985).

4.3.4.2. The Ring

The ring is one of the basic and earliest network connections. With n nodes, they are ordered linearly so that every node has a unique label I wherein where 0 I n – 1. Each node is linked to its left and right neighbors. Therefore, a node identified as I is linked to the nodes identified as labeled i + 1mod n and i – 1mod n (see Figure 4.9). The ring is used by devices with local memory (LMMs) (Kahle et al., 2005; Jin et al., 2008).

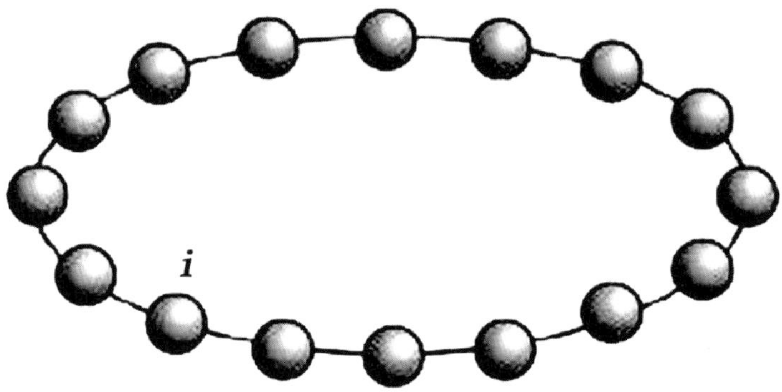

Figure 4.9. A ring – each node represents a processor unit with local memory.

Source: https://www.amazon.com/Introduction-Parallel-Computing-Ananth-Grama/dp/0201648652.

4.3.4.3. 2D-Mesh

A two-dimensional grid is a connection system that may be formed in a rectangle, for each switch having a unique label i, j), where (i, j), where 0 i X – 1 and 0 j Y – 1 (see Figure 4.10). X and Y variables dictate the sizes of the mesh's sides. Consequently, XY is the amount of switching in a mesh. Excluding the switches on the edges of the mesh, each switch has six neighbors: one from the north, one from the south, one from the east, as well as one from the west. The switch where $0 < i < X - 1$ and $0 < j < Y - 1$, is connected to the switches labeled as (i, j + 1), (i, j – 1), (i + 1, j), and (i – 1, j) (Barr et al., 2005).

Meshes are often seen in local-memory machines (LMMs): a processor (together with its memory location) is attached to every switch, and distant memory requests are accomplished by exchanging packets across the mesh (Figure 4.11) (Pashler, 1990).

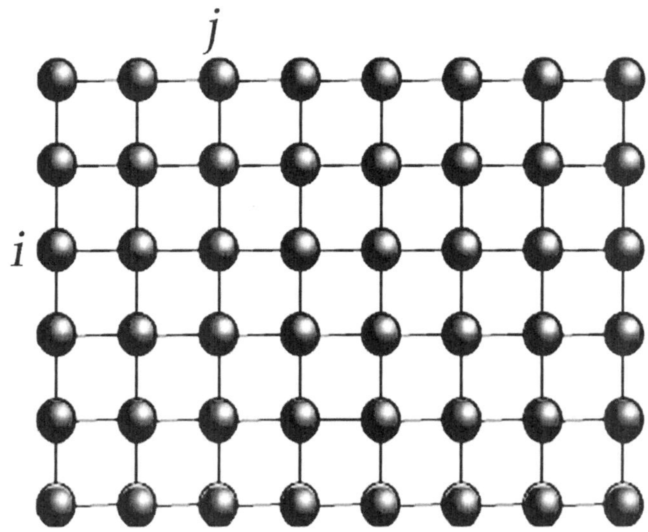

Figure 4.10. *A 2D-mesh – every j node contains a local memory-equipped processing unit.*

Source: https://link.springer.com/book/10.1007/978-3-319-98833-7.

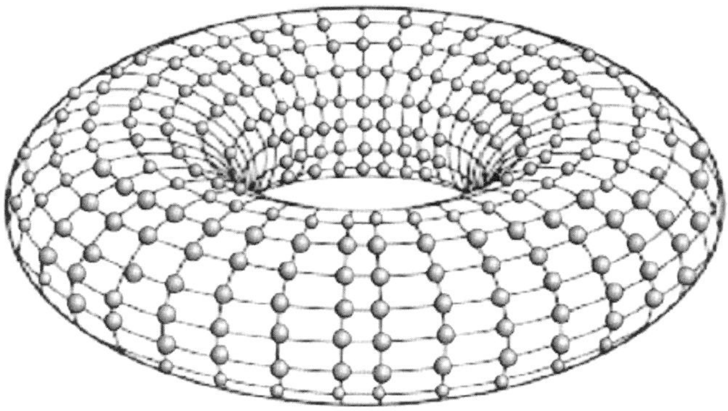

Figure 4.11. A 2D torus – each node represents a local memory-equipped processor.

Source: https://link.springer.com/book/10.1007/978-3-319-98833-7.

4.3.4.4. 3D-Mesh and 3D-Torus

Likewise, a cube is a 3D mesh (see Figure 4.12). Currently, every switch in a mesh does have a unique label i, j, k), where (i, j, k), where 0 i X − 1, 0 j Y − 1, and 0 k Z − 1. The dimensions of the edges of the mesh are determined by the numbers X, Y, and Z, hence its switching frequency is XY Z. Each switch is now linked to six neighbor nodes: one from the north, one from the south, one from the east, one from the west, one upwards, and one downwards (Blake et al., 1988).

A 3D-mesh may be expanded into a toroidal 3D-mesh by introducing sides that link nodes on opposite teams of the 3D-mesh. (Image omitted.) A switch labeled (i, j, k) is linked to the switches (i + 1mod X, j,k),(i − 1mod X, j,k), (i, j + 1mod Y,k), (i, j − 1mod Y,k), (i, j,k + 1mod Z) and (i, j,k − 1mod Z). Local-memory machines make use of 3D-meshes and toroidal 3D-meshes (LMMs) (McLeod, 1977).

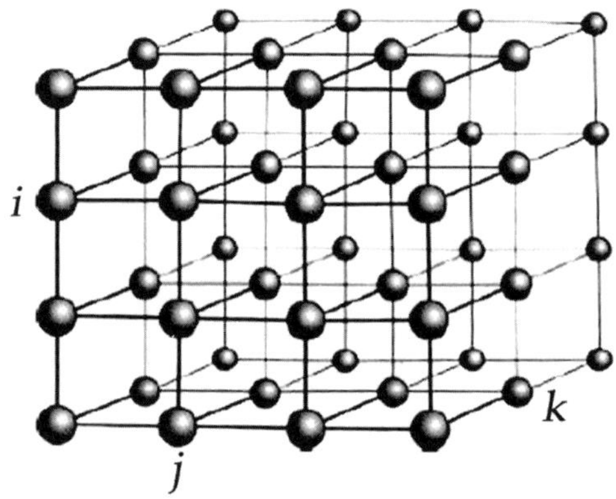

Figure 4.12. A 3D-mesh – every node contains a local memory-equipped processor.

Source: https://link.springer.com/book/10.1007/978-3-319-98833-7.

4.3.4.5. Hypercube

For certain b 0, a hypercube is an interconnection system with n = 2b units (see Figure 4.13). Every node has its label, which is made out of b bits. A data connection connects two nodes and only then if their names vary in exactly one-bit position. As a result, every hypercube component has b = log2 n neighbors. Hypercubes are employed in devices with local memory (LMMs) (Alglave et al., 2009).

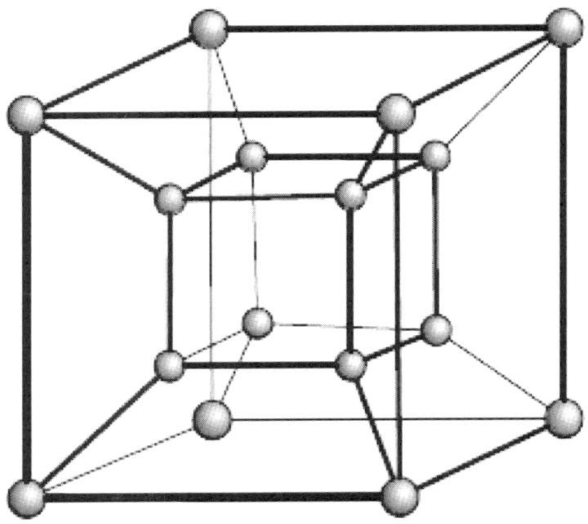

Figure 4.13. A hypercube – each node represents a local memory-equipped processor.

Source: https://link.springer.com/book/10.1007/978-3-319-98833-7.

4.3.4.6. The k-ary d-Cube Family of Network Topologies

Surprisingly, the ring, 2D-torus, 3D-torus, hypercube, and several other topologies are all members of the same family circle of k-ary d-cube topologies (Poplavko et al., 2003).

Assuming k 1 and d 1, the k-ary d-cube topology is a collection of some "gridlike" topologies which are built in the same way. To put it another way, the k-ary d-cube topology is an extension of existing topologies. The variable d is referred to as the component of such topologies, while k is the set of branches along with every one of the d directions. The construction of the k-ary d-cube topology is described experimentally (on the dimension d): A k-ary d-cube is built using k additional k-ary (d 1)-cubes by linking identically positioned terminals into circles (Archibald and Baer, 1986).

We can use this inductive concept to build genuine k-ary d-cube topologies and study their topological as well as designs incorporated (Lee et al., 2008).

4.3.4.7. Multistage Network

A network with many stages' links one group of switches, the input switch mode, to another group, the output switches. The system does this via a series of phases, which are composed of relays (see Figure 4.14). Specifically, the input switches constitute the initial phase as well as the output switching in the final phase. The number of layers is referred to as the network's depth. A multiple-stage network typically enables the data transmission from every source switching to every output switching. It is accomplished through a route that spans all system phases from 1 to d in sequential order. There are several distinct topologies for multistage networks (Leutenegger and Vernon, 1990).

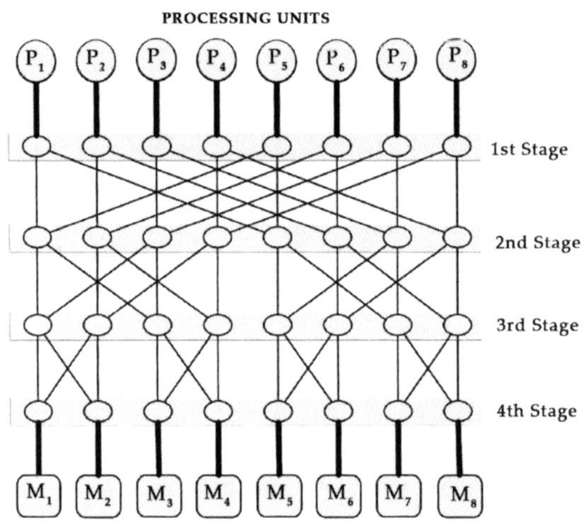

Figure 4.14. A four-stage interconnection network that can link eight CPU units to eight system memory. Every switch can link a unique pairing of inlet and outlet ports.

Source: https://www.cs.purdue.edu/homes/ayg/book/Slides/.

Memory-module machines (MMMs) often use multistage systems, where processor modules are connected to input switching and main memory are connected to outlet switches (Wolf et al., 2008).

4.3.4.8. Fat Tree

A fat tree is a system depending on the framework of a tree (see Figure 4.15). In a fat tree, therefore, the branches closer to the tree's base are "fatter" (thicker) than those farther from the tree's core. Every component of a fat tree is supposed to show several network switches, while every edge represents multiple communication lines. The greater an edge's capability and the more pathways it covers, the greater its width. Therefore, the capabilities of the borders at the base of the fat-tree are much more than those around the branches (Ramamritham et al., 1990).

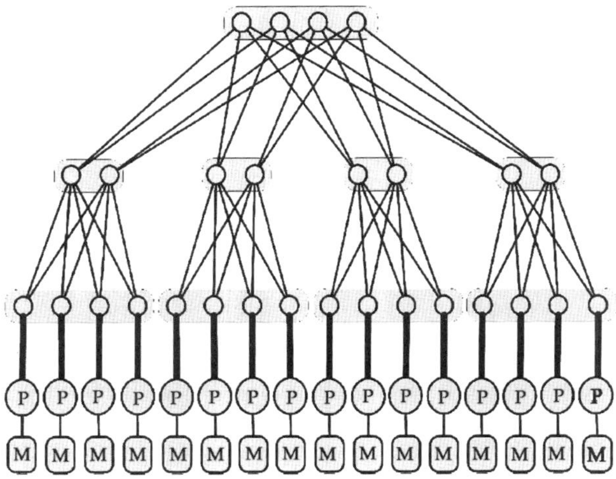

Figure 4.15. A hefty tree – each switch is capable of connecting any two event channels.

Source: https://www.cs.purdue.edu/homes/ayg/book/Slides/.

Fat plants can be utilized to build local-memory machines (LMMs): processing elements, as well as their local memories, are linked to the leaf of the fat tree so a text through one processing system to some other first moves up the tree to the lowest single origin of the two processing elements as well as then bottom the tree to the location processing system (Paulin et al., 2006).

4.4. PARALLEL COMPUTATIONAL COMPLEXITY

To analyze the difficulty of computing issues and associated parallel methods, we need new fundamental concepts. We shall now present many of them (Maeda et al., 2016).

4.4.1. Problem Instances and Their Sizes

Let us represent a computing issue. Typically, we encounter a single case of the issue in practice. The instance is produced by substituting real data for the parameters in the specification. Because this may be accomplished in several ways, each result in a unique instance, the issue can be seen as a collection of all potential examples of π. To every case, a natural integer, which we name the length of the case and indicate by, may be assigned (Meier et al., 2003).

- **size(π):** Casually, size() is the approximate sufficient space required to express in a computer-accessible format; in fact, size() comes down to the problem Π. Why then do we require case sizes? When evaluating the speed of an algorithm A solving an issue Π, we often like to understand how A's runtime varies with the set of input cases of Π. Specifically, we would like to identify a function (Wang and Xiao, 2018).
- **T(n):** Where the value at n represents the completion time of A when applied to cases of size n. In reality, we are mainly concerned with the growth rate of T(n), or how rapidly T(n) expands as n increases (Schliecker et al., 2008).

For example, if T(n) = n, then A's runtime is a linear function of n; hence, if the amount of the test problems is doubled, so is A's reaction time. More broadly, if T(n) = nconst(const 1), then A's runtime is a linear transformation of n; if the number of the test problems is doubled, then A's runtime is multiplied by 2 const. If, however, we find that $T(n) = 2^n$, which is an exponential function of n, then things become dramatic: doubling the size n of problem instances causes A to run 2^n-times longer! So, doubling the size from 10 to 20 and then to 40 and 80, the execution time of A increases 2^{10} (≈thousand) times, then 2^{20} (≈million) times, and finally 2^{40} (≈thousand billion) times (Levin et al., 2010).

4.4.2. Number of Processing Units versus Size of Problem Instances

On a computer C(p) with p processing elements, we described the parallel completion time Tpar, speed-up S, and effectiveness E of a parallel program P for finding a solution. Allow us to extend such definitions to include the number n of the cases of Π (Zeng et al., 2008). We remove the appropriate indices to clarify the syntax since the code P for resolving and the computer C(p) are implicitly known. The parallel run time Tpar(n), the speedup S(n), and the efficiency E(n) of solving's instances Π. of size n are obtained as follows: def Tseq(n) S(n) (Brekling et al., 2008):

$$S(n) \stackrel{def}{=} \frac{T_{seq}(n)}{T_{par}(n)},$$

$$E(n) \stackrel{def}{=} \frac{S(n)}{p}.$$

So, let us choose a random *n* and assume we are only capable of resolving cases of whose size is *n*.

However, if C(p) has several processing units, i.e., p is just too little, the program P's theoretical concurrency would not be completely utilized throughout the execution on C(p), resulting in a low speedup S(n) of P (Dutta et al., 2016). Similarly, if C(p) contains an excessive number of production units, i.e., p is too big, part of the processing elements will remain idle throughout the running of the program P, resulting in a slow speedup of P. This poses a critical issue, which merits additional thought: How much processing units should C(p) contain such that the performance of P is maximized for all occurrences of size n (Marsan and Gerla, 1982)?

The response is likely to be influenced by the kind of C, that is, the multiprocessor design that underpins the parallel computer C. We may not even be capable to get effective solutions to this question till we adopt the multiprocessor design. Nonetheless, we may draw certain broad conclusions that apply to any form of C (Zeng et al., 2008). First, note that we should let n increase, then p must expand as well; else, p would ultimately are becoming too tiny in comparison to n, rendering C(p) unable of fully harnessing P's parallelism capability. As a result, we may consider p, the number of computational units required to achieve maximum performance, to be a factor of n, the magnitude of the prediction problem in question. Intuition and experience teach us that a bigger example of an issue necessitates at least the same number of computational units as a lesser one. To summarize, we may (Sengupta and Dahbura, 1992):

$$p = f(n)$$

wherein; f: N N denotes a nondecreasing function, e.g., f (n) f (n + 1) for all *n*.

Second, consider how rapidly f (n) may expand as n grows. Assume that f (n) increases at an exponential rate. Scientists have shown that when a parallel system has an increasingly large number of computational elements, there must be lengthy communication links between some of them. As the distance between certain connecting processor units increases larger, the connection periods among them grow in lockstep, blemishing the potentially attainable speed-up. All of this is based on our real-world, 3D space, since (Martin et al., 2005):

- Every processing element as well as data connection takes up that space that is not zero;
- The lowest sphere with infinitely numerous processing elements plus communications networks has exponential size.

In conclusion, having an exponential amount of processing elements is unrealistic and might lead to potentially difficult circumstances.

Assume that f (n) increases polynomially, that is, that f is a linear transformation of n. If poly(n) and exp(n) are polynomial as well as exponential functions, correspondingly, calculus informs us there is a n > 0 such that poly(n) exp(n) for every n > n; that really is, poly(n) is ultimately controlled by exp(n) (n). In other terms, a polynomial function poly(n) develops shorter monotonically than an exponent function exp (n). Poly(n) and exp(n) are two random functions of n to be aware of (Enslow, 1977).

As a result, we obtain f (n) = poly(n), and the number of processing units is:

$p = poly(n)$

n is a mathematical expression, and poly(n) is a polynomial function of n. Polynomial functions with "irrationally" high degrees, such as n100, are implicitly rejected here. We are looking for considerably lower degrees, such as 2, 3, or 4, that will result in more practical as well as cheap CPU numbers p.

In conclusion, we have received an answer to the question above that falls short of our expectations owing to the generalization of C and Π, as well as constraints imposed by nature and economics. Regardless, the solution indicates that p must be a moderate-degree polynomial function of n (Tsuei and Yamamoto, 2003).

4.4.3. The Class NC of Efficiently Parallelizable Problems

Let P be a problem-solving algorithm on CRCW-PRAM (p). P would be executed at most O(log p) times shorter on EREW-PRAM(p) than on CRCW-PRAM (p). Let us all utilize the preceding section's findings as well as assume that p = poly (n). As a result, log p = logpoly(n) = O. (logn) (Moller and Tofts, 1990).

When paired with Theorem 2.1, that suggests that for p = poly(n), the implementation of P on EREW-PRAM(p) would be no more than O(logn) times less than on CRCW-PRAM (p) (Barr et al., 2005).

However, for p = poly(n), selecting modeling from the categories CRCW-PRAM(p), CREW-PRAM(p), and EREW-PRAM(p) to run programming impacts the program's completion time by a ratio of the order O(logn), wherein n is the number of issue cases to be addressed. In other terms, the runtime of a program does not change much depending on the variation of PRAM used to execute it (Jacopini and Sontacchi, 1990).

It drives us to define a class of computing issues that includes all issues with "rapid" parallel methods that need an "appropriate" amount of processing elements. However, what exactly do the terms "quick" and "fair" imply? We saw in the last part that if the quantity of processing elements is polynomial in '*n*,' it is acceptable. In terms of definition, a parallel algorithm is deemed quick if its parallel rum duration is polylogarithmic in n. That is great, and what does "polylogarithmic" imply now? Here is an explanation (Beggs and Tucker, 2006).

- **Definition 1:** A polynomial is polylogarithmic in *n* if this is linear in logn, that is, if for any k 1 it is ak(logn)k + ak1(logn)k1 + + a1(logn)1 + a0.

To prevent parenthesis clumping, we commonly use logi n rather than (logn)i.

Beyond O, the summation aklogkn + ak1logk1n + + a0 is monotonically limited (logk n).

We are now prepared to officially present the class of issues in which we are concerned.

- **Definition 2:** Let NC denote the category of computing questions that can be answered in polylogarithmic duration on PRAM with a polynomial series of processing elements.

If an issue Π belongs to the NC class, it can be solved in polylogarithmic parallel time using polynomially numerous processing elements, independent of the kind of PRAM utilized. In other terms, the NC class is resistant to fluctuations in PRAM. How do we know this? If we swap one PRAM variation for another, the parallel runtime O(logk n) could only grow by a ratio of O(logn) to O(logk+1 n), and that's still polylogarithmic (Hu et al., 1998).

- **Exemplification 2.1:** Assume we are assigned the issue "add n supplied numbers." Then "add numbers 10, 20, 30, 40, 50, 60, 70, 80" is a size() = 8case of the issue. And let us concentrate on all occurrences of size 8, that is, cases of the kind $\pi \equiv$ "additional integers a_1, a_2, a_3, a_4, a_5, a_6, a_7, a8" (Fischer et al., 2003).

Tseq(8) = 7 steps are required to compute the total $a_1 + a_2 + a_3 + a_4 + a_5 + a_6 + a_7 + a_8$ among every step increasing a next integer to the combination of the preceding ones (Abdel-Jabbar et al., 1998).

Therefore, the integers a_1, a_2, a_3, a_4, a_5, a_6, a_7, a_8 may be added together in Tpar(8) = 3 parallel steps employing = 4 processing elements that interact in a trec-like arrangement, as shown in Figure 4.16. So, every processing element keeps adding two adjoining input numbers in the first phase, s = 1 (Li and Abushaikha, 2021). In the following steps, s 2, two adjoining last half results are combined to make a new, combined incomplete result. The above tree-like incorporation of preliminary results keeps going till the 2s + 1 > 8. Step s = 1 involves all 4 processing elements in arithmetic; step s = 2 involves two processing elements (P_3 and P_4) idling; and step s = 3 contains three processing elements (P_2, P_3, and P_4) idling (Feldman and Shapiro, 1992).

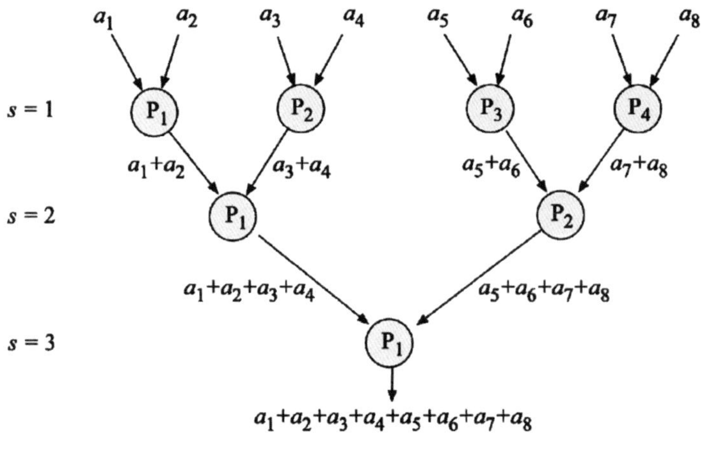

Figure 4.16. Parallel addition of eight digits using four processing elements.

Source: https://www.cs.purdue.edu/homes/ayg/book/Slides/.

Overall, instances (n) could be fixed concurrently. Tpar = logn =O(logn), with computer processors interacting in a tree-like pattern.

$\frac{n}{2}\rceil = O(n)$

∈ $= T^{seq}par(n) =$ ^{n} *terns.*

Hence, Π NC and the associated speedup is:

S(n) T (n) O(logn).

It is worth noting that the effectiveness of the tree-like concurrent accumulation of n numbers in the previous sentence is relatively poor, E(n) = O. (log1 n). The explanation here is evident: only 50% of the processing elements involved in a concurrent phase s would be involved in the following parallel phase s + 1, whereas the remaining computer processors would be idle till the calculation is completed. Brent's Theorem would resolve these issues in the following part (Smith, 1993).

4.5. LAWS AND THEOREMS OF PARALLEL COMPUTATION

Brent's theorem, which helps predict the bottom limit on the number of processing elements required to maintain a particular parallel computation time, is described as follows. And we will look at Amdahl's law, which is utilized to determine the theoretical increase in speed of a parallel system with many portions that allow for varying speeds (Bandini et al., 2001).

4.5.1. Brent's Theorem

Whenever the amount of processing elements is lowered, Brent's theorem allows us to measure the effectiveness of a parallel program (Zhang et al., 2019).

Allow M to be a PRAM of any kind with an undefined series of processing units. We suppose, more explicitly, that the amount of processing elements has always been adequate to meet all of the requirements of any parallel application (Marszalek and Podhaisky, 2016).

When a parallel program *P* is run on *M*, different numbers of operations of *P* are performed, at each step, by different processing units of *M*. Suppose that a total of

W

operations are performed during the parallel execution of *P* on *M* (*W* is also called the **work** of *P*), and denote the parallel runtime of *P* on *M* by

$T_{par,}$ M(P).

Let us now reduce the number of processing units of *M* to some fixed number

p

and denote the obtained machine with the reduced number of processing units by *R*.

(Revol and Théveny, 2014).

Applications of Brent's Theorem: Brent's Theorem has implications whenever we wish to decrease the number of processing elements whenever feasible while maintaining parallel time complications. With O(n) processing elements, we could add n numbers in O(logn) time. Is it possible to achieve a similar thing with asymptotically fewer processing elements? Yes, certainly. According to Brent's Theorem, O(n/logn) computer processors are sufficient to add n integers in O(logn) parallel duration (Moore, 1997).

4.5.2. Amdahl's Law

Inherently, we might anticipate that twice the amount of processing elements would half the parallel runtime and that twice the number of processing units a second time might further reduce the parallel runtime. In plenty of other words, we anticipate that the acceleration from parallelization is proportional to the number of processing elements (see Figure 4.17) (Baghdadi et al., 2001).

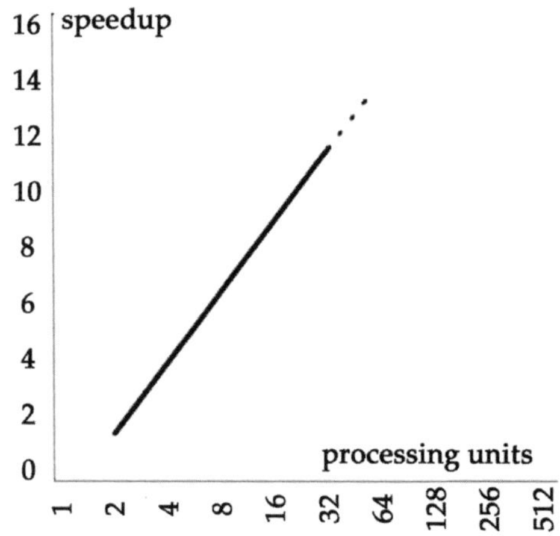

Figure 4.17. Estimated (linear) greatly accelerate as a result of the number of processing units.

Source: https://www.lrde.epita.fr/~ricou/intro_parallel_comp.pdfs.

Therefore, linear acceleration from parallelization is only an idea that is unlikely to be realized. Several parallel algorithms do this in practice. The majority of parallel programs exhibit a near-linear speed-up for tiny groups of processing units, that straightens to a fixed rate for massive quantities of processing components (see Figure 4.18) (Bodirsky et al., 2007).

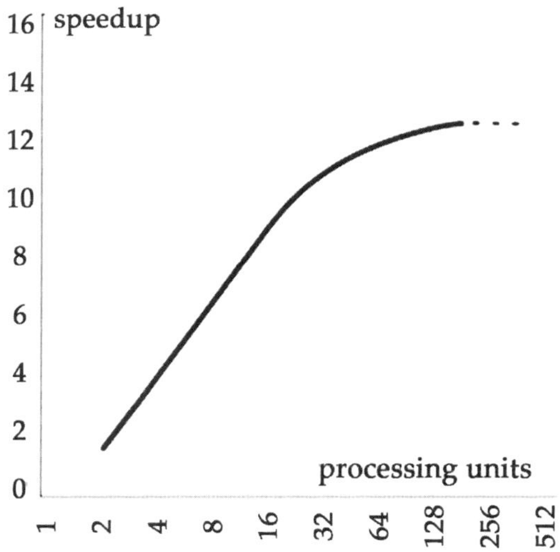

Figure 4.18. Real increase as a result of the number of processing elements.

Source: https://www.lrde.epita.fr/~ricou/intro_parallel_comp.pdfs.

Setting the Stage: Could this unexpected behavior be explained? The solution would be deduced from two straightforward instances (Mishra et al., 2012).

Suppose P is a serial application that reads files from such a disk as follows:

- *P is a two-part sequence,* $P = P_1P_2$*;*
- P_1 *examines the disk's directory, generates a collection of data files, and passes the list to* P_2*;*
- P_2 forwards each file on the collection to the main processor for additional analysis.

Searching the disk drive is an inherently periodic activity that cannot be accelerated by introducing more processing units to P_1. In comparison, P_2 may be accelerated by introducing more processing elements; for instance, every file could be sent to its processing system. In conclusion, a serial program may be considered as a series of two sections that vary in their parallelizability, or parallel processing capability (Sadegh, 1993).

Let P be the same as above. Assuming the (sequence) implementation of P requires 20 minutes, the preceding is true (see Figure 4.19). The non-parallelizable P_1 operates for 2 minutes, whereas the parallel processing P_2 runs for 18 minutes (Goldschlager and Parberry, 1986).

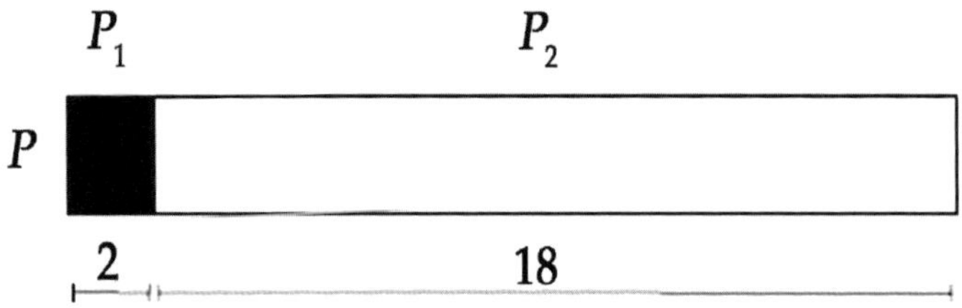

Figure 4.19. Non-parallelizable P is composed of a parallelizable P_1 and a parallelizable P_2. On a single P processing unit, P_1 operates for 2 minutes and P_2 for 18 minutes.

Source: https://www.lrde.epita.fr/~ricou/intro_parallel_comp.pdfs.

Since only P_2 can support extra processing elements, the parallel execution period Tseq(P) of the entire P could not be less than the time Tseq(P_1) collected by the non-parallelizable component P_1 (that is, 2 minutes), irrespective of the number of extra processing elements involved in the parallel implementation of P (Ghorbel et al., 2000). In conclusion, if portions of a sequential circuit vary in their possible parallelisms, they would also vary in their possible acceleration from the additional series of processing units, hence the speed increase of the entire program would rely on the sequential operation run of the individual components (Figure 4.20) (Keerthi and Gilbert, 1988).

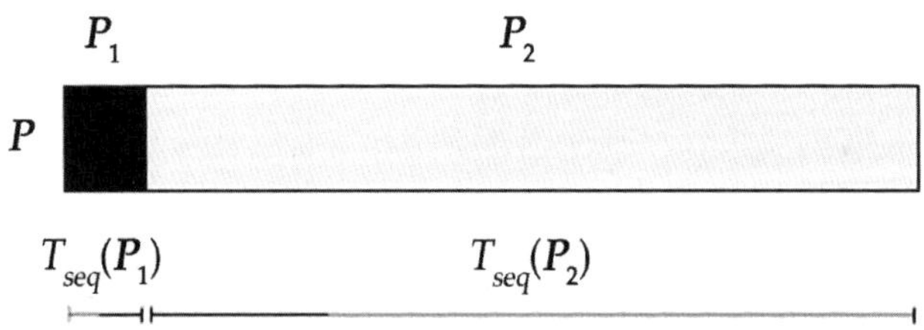

Figure 4.20. P comprises P_1, which cannot be parallelized, and P_2, which can be parallelized. P_1 needs Tseq(P_1) time, as well as P_2, needs Tseq(P_2) time to finish on a particular processing system.

Source: https://www.lrde.epita.fr/~ricou/intro_parallel_comp.pdfs.

The hints that the preceding cases illuminated are summarized here. In principle, a parallel computer may divide a program P into two pieces (Devine and Flaherty, 1996):

- Component P_1 that does not get an advantage from many processing elements; and
- Portion P_2 that is aided by several processing elements;

In addition to P_2's advantage, the sequential runtime of P_1 and P_2 affect the parallel completion time of the entire P (as well as, hence, P's speedup).

REFERENCES

1. Abdel-Jabbar, N., Kravaris, C., & Carnahan, B., (1998). A partially decentralized state observer and its parallel computer implementation. *Industrial & Engineering Chemistry Research, 37*(7), 2741–2760.
2. Alglave, J., Fox, A., Ishtiaq, S., Myreen, M. O., Sarkar, S., Sewell, P., & Nardelli, F. Z., (2009). The semantics of Power and ARM multiprocessor machine code. In: *Proceedings of the 4th Workshop on Declarative Aspects of Multicore Programming* (Vol. 1, pp. 13–24).
3. Andrews, J. B., & Polychronopoulos, C. D., (1991). An analytical approach to performance/cost modeling of parallel computers. *Journal of Parallel and Distributed Computing, 12*(4), 343–356.
4. Archibald, J., & Baer, J. L., (1986). Cache coherence protocols: Evaluation using a multiprocessor simulation model. *ACM Transactions on Computer Systems (TOCS), 4*(4), 273–298.
5. Arge, L., Goodrich, M. T., Nelson, M., & Sitchinava, N., (2008). Fundamental parallel algorithms for private-cache chip multiprocessors. In: *Proceedings of the Twentieth Annual Symposium on Parallelism in Algorithms and Architectures* (Vol. 1, pp. 197–206).
6. Asanovic, K., Bodik, R., Demmel, J., Keaveny, T., Keutzer, K., Kubiatowicz, J., & Yelick, K., (2009). A view of the parallel computing landscape. *Communications of the ACM, 52*(10), 56–67.
7. Axelrod, T. S., (1986). Effects of synchronization barriers on multiprocessor performance. *Parallel Computing, 3*(2), 129–140.
8. Bader, D. A., & Cong, G., (2005). A fast, parallel spanning tree algorithm for symmetric multiprocessors (SMPs). *Journal of Parallel and Distributed Computing, 65*(9), 994–1006.
9. Baghdadi, A., Lyonnard, D., Zergainoh, N. E., & Jerraya, A. A., (2001). An efficient architecture model for systematic design of application-specific multiprocessor SoC. In: *Proceedings Design, Automation and Test in Europe: Conference and Exhibition 2001* (pp. 55–62). IEEE.
10. Bandini, S., Mauri, G., & Serra, R., (2001). Cellular automata: From a theoretical parallel computational model to its application to complex systems. *Parallel Computing, 27*(5), 539–553.
11. Barney, B., (2010). *Introduction to Parallel Computing* (Vol. 6, No. 13, p. 10). Lawrence Livermore National Laboratory.
12. Barr, K. C., Pan, H., Zhang, M., & Asanovic, K., (2005). Accelerating multiprocessor simulation with a memory timestamp record. In: *IEEE*

International Symposium on Performance Analysis of Systems and Software, 2005; ISPASS 2005 (Vol. 1, pp. 66–77). IEEE.

13. Beggs, E. J., & Tucker, J. V., (2006). Embedding infinitely parallel computation in Newtonian kinematics. *Applied Mathematics and Computation, 178*(1), 25–43.
14. Benini, L., Bertozzi, D., Bogliolo, A., Menichelli, F., & Olivieri, M., (2005). Mparm: Exploring the multi-processor SoC design space with systemC. *Journal of VLSI Signal Processing Systems for Signal, Image and Video Technology, 41*(2), 169–182.
15. Bhandarkar, S. M., & Arabnia, H. R., (1995). The REFINE multiprocessor—Theoretical properties and algorithms. *Parallel Computing, 21*(11), 1783–1805.
16. Blake, J. T., Reibman, A. L., & Trivedi, K. S., (1988). Sensitivity analysis of reliability and performability measures for multiprocessor systems. In: *Proceedings of the 1988 ACM SIGMETRICS Conference on Measurement and Modeling of Computer Systems* (Vol. 1, pp. 177–186).
17. Bodirsky, M., Giménez, O., Kang, M., & Noy, M., (2007). Enumeration and limit laws for series–parallel graphs. *European Journal of Combinatorics, 28*(8), 2091–2105.
18. Bracker, S., Gounder, K., Hendrix, K., & Summers, D., (1996). A simple multiprocessor management system for event-parallel computing. *IEEE Transactions on Nuclear Science, 43*(5), 2457–2464.
19. Brekling, A., Hansen, M. R., & Madsen, J., (2008). Models and formal verification of multiprocessor system-on-chips. *The Journal of Logic and Algebraic Programming, 77*(1, 2), 1–19.
20. Corcoran, A. L., & Wainwright, R. L., (1994). A parallel island model genetic algorithm for the multiprocessor scheduling problem. In: *Proceedings of the 1994 ACM Symposium on Applied Computing* (Vol. 1, pp. 483–487).
21. Covington, R. G., Dwarkadas, S., Jump, J. R., Sinclair, J. B., & Madala, S., (1991). The efficient simulation of parallel computer systems. *International Journal in Computer Simulation, 1*(1), 31–58.
22. Culler, D. E., Karp, R. M., Patterson, D., Sahay, A., Santos, E. E., Schauser, K. E., & Von, E. T., (1996). LogP: A practical model of parallel computation. *Communications of the ACM, 39*(11), 78–85.

23. Culler, D., Karp, R., Patterson, D., Sahay, A., Schauser, K. E., Santos, E., & Von, E. T., (1993). LogP: Towards a realistic model of parallel computation. In: *Proceedings of the Fourth ACM SIGPLAN Symposium on Principles and Practice of Parallel Programming* (Vol. 1, pp. 1–12).
24. Dennis, J. B., (1994). Machines and models for parallel computing. *International Journal of Parallel Programming, 22*(1), 47–77.
25. Devine, K. D., & Flaherty, J. E., (1996). Parallel adaptive hp-refinement techniques for conservation laws. *Applied Numerical Mathematics, 20*(4), 367–386.
26. Dhillon, I. S., & Modha, D. S., (2002). A data-clustering algorithm on distributed memory multiprocessors. In: *Large-Scale Parallel Data Mining* (Vol. 1, pp. 245–260). Springer, Berlin, Heidelberg.
27. Dongarra, J., Foster, I., Fox, G., Gropp, W., Kennedy, K., Torczon, L., & White, A., (2003). *Sourcebook of Parallel Computing* (Vol. 3003, pp. 3–9). San Francisco: Morgan Kaufmann Publishers.
28. Drozdowski, M., (1996). Scheduling multiprocessor tasks—An overview. *European Journal of Operational Research, 94*(2), 215–230.
29. Drozdowski, M., (1997). *Selected Problems of Scheduling Tasks in Multiprocessor Computer Systems* (Vol. 1, pp. 2–8). Politechnika Poznańska.
30. Dutta, S., Cadambe, V., & Grover, P., (2016). Short-dot: Computing large linear transforms distributedly using coded short dot products. *Advances In Neural Information Processing Systems, 1*, 29.
31. Enslow, Jr. P., (1977). Multiprocessor organization—A survey. *ACM Computing Surveys (CSUR), 9*(1), 103–129.
32. Ezhova, N. A., & Sokolinskii, L. B., (2018). *Parallel Computational Model for Multiprocessor Systems with Distributed Memory* (Vol. 7, No. 2, p. 32–49). Bulletin of the South Ural State University. Series "Computational Mathematics and Informatics".
33. Ezhova, N. A., & Sokolinskii, L. B., (2019). *Survey of Parallel Computation Models* (Vol. 8, No. 3, pp. 58–91). Bulletin of the South Ural State University. Series "Computational Mathematics and Informatics".
34. Feldman, Y., & Shapiro, E., (1992). Spatial machines: A more realistic approach to parallel computation. *Communications of the ACM, 35*(10), 60–73.

35. Fischer, J., Gorlatch, S., & Bischof, H., (2003). Foundations of data-parallel skeletons. In: *Patterns and Skeletons for Parallel and Distributed Computing* (Vol. 1, pp. 1–27). Springer, London.
36. Francis, R. S., & Pannan, L. J., (1992). A parallel partition for enhanced parallel quicksort. *Parallel Computing, 18*(5), 543–550.
37. Gajski, D. D., & Peir, J. K., (1985). Essential issues in multiprocessor systems. *Computer, 18*(06), 9–27.
38. Garland, M., Le Grand, S., Nickolls, J., Anderson, J., Hardwick, J., Morton, S., & Volkov, V., (2008). Parallel computing experiences with CUDA. *IEEE Micro, 28*(4), 13–27.
39. Ghorbel, F. H., Chételat, O., Gunawardana, R., & Longchamp, R., (2000). Modeling and set point control of closed-chain mechanisms: Theory and experiment. *IEEE Transactions on Control Systems Technology, 8*(5), 801–815.
40. Goldschlager, L. M., & Parberry, I., (1986). On the construction of parallel computers from various bases of Boolean functions. *Theoretical Computer Science, 43*, 43–58.
41. Grindley, R., Abdelrahman, T., Brown, S., Caranci, S., DeVries, D., Gamsa, B., & Zilic, Z., (2000). The NUMAchine multiprocessor. In: *Proceedings 2000 International Conference on Parallel Processing* (Vol. 1, pp. 487–496). IEEE.
42. Gropp, W. W., & Lusk, E. L., (1995). A taxonomy of programming models for symmetric multiprocessors and SMP clusters. In: *Programming Models for Massively Parallel Computers* (Vol. 1, pp. 2–7). IEEE.
43. Guan, N., Gu, Z., Deng, Q., Gao, S., & Yu, G., (2007). Exact schedulability analysis for static-priority global multiprocessor scheduling using model-checking. In: *IFIP International Workshop on Software Technologies for Embedded and Ubiquitous Systems* (Vol. 1, pp. 263–272). Springer, Berlin, Heidelberg.
44. Guan, Y., Xiao, W. Q., Cheung, R. K., & Li, C. L., (2002). A multiprocessor task scheduling model for berth allocation: Heuristic and worst-case analysis. *Operations Research Letters, 30*(5), 343–350.
45. Hu, Z., Takeichi, M., & Chin, W. N., (1998). Parallelization in calculational forms. In: *Proceedings of the 25th ACM SIGPLAN-SIGACT Symposium on Principles of Programming Languages* (Vol. 1, pp. 316–328).

46. Jacopini, G., & Sontacchi, G., (1990). Reversible parallel computation: An evolving space-model. *Theoretical Computer Science, 73*(1), 1–46.
47. Jin, X., Furber, S. B., & Woods, J. V., (2008). Efficient modelling of spiking neural networks on a scalable chip multiprocessor. In: *2008 IEEE International Joint Conference on Neural Networks (IEEE World Congress on Computational Intelligence)* (Vol. 1, pp. 2812–2819). IEEE.
48. Kahle, J. A., Day, M. N., Hofstee, H. P., Johns, C. R., Maeurer, T. R., & Shippy, D., (2005). Introduction to the cell multiprocessor. *IBM journal of Research and Development, 49*(4, 5), 589–604.
49. Keerthi, S. S., & Gilbert, E. G., (1988). Optimal infinite-horizon feedback laws for a general class of constrained discrete-time systems: Stability and moving-horizon approximations. *Journal of Optimization Theory and Applications, 57*(2), 265–293.
50. Kessler, C., & Keller, J., (2007). *Models for Parallel Computing: Review and Perspectives* (Vol. 24, pp. 13–29). Communications society for computer science eV, Parallel algorithms and computer structures.
51. Khan, M. M., Lester, D. R., Plana, L. A., Rast, A., Jin, X., Painkras, E., & Furber, S. B., (2008). SpiNNaker: Mapping neural networks onto a massively-parallel chip multiprocessor. In: *2008 IEEE International Joint Conference on Neural Networks (IEEE World Congress on Computational Intelligence)* (Vol. 1, pp. 2849–2856). IEEE.
52. Kistler, M., Perrone, M., & Petrini, F., (2006). Cell multiprocessor communication network: Built for speed. *IEEE Micro, 26*(3), 10–23.
53. Krämer, O., & Mühlenbein, H., (1989). Mapping strategies in message-based multiprocessor systems. *Parallel Computing, 9*(2), 213–225.
54. Kurgalin, S., & Borzunov, S., (2019). *A Practical Approach to High-Performance Computing* (Vol. 206, pp. 2–9). Springer International Publishing.
55. Lee, B. C., Collins, J., Wang, H., & Brooks, D., (2008). CPR: Composable performance regression for scalable multiprocessor models. In: *2008 41st IEEE/ACM International Symposium on Microarchitecture* (Vol. 1, pp. 270–281). IEEE.
56. Leutenegger, S. T., & Vernon, M. K., (1990). The performance of multiprogrammed multiprocessor scheduling algorithms. In: *Proceedings of the 1990 ACM SIGMETRICS Conference on*

Measurement and Modeling of Computer Systems (Vol. 1, pp. 226–236).

57. Levin, G., Funk, S., Sadowski, C., Pye, I., & Brandt, S., (2010). DP-FAIR: A simple model for understanding optimal multiprocessor scheduling. In: *2010 22nd Euromicro Conference on Real-Time Systems* (Vol. 1, pp. 3–13). IEEE.
58. Li, J., & Martinez, J. F., (2005). Power-performance considerations of parallel computing on chip multiprocessors. *ACM Transactions on Architecture and Code Optimization (TACO), 2*(4), 397–422.
59. Li, J., & Martinez, J. F., (2005). Power-performance implications of thread-level parallelism on chip multiprocessors. In: *IEEE International Symposium on Performance Analysis of Systems and Software, 2005: ISPASS 2005* (Vol. 1, pp. 124–134). IEEE.
60. Li, J., & Martinez, J. F., (2006). Dynamic power-performance adaptation of parallel computation on chip multiprocessors. In: *The Twelfth International Symposium on High-Performance Computer Architecture, 2006* (Vol. 1, pp. 77–87). IEEE.
61. Li, K., (1988). IVY: A shared virtual memory system for parallel computing. *ICPP*, *88*(2), 94.
62. Li, L., & Abushaikha, A., (2021). A fully-implicit parallel framework for complex reservoir simulation with mimetic finite difference discretization and operator-based linearization. *Computational Geosciences, 1*, 1–17.
63. Li, Y., & Zhang, Z., (2018). Parallel computing: Review and perspective. In: *2018 5th International Conference on Information Science and Control Engineering (ICISCE)* (Vol. 1, pp. 365–369). IEEE.
64. Lin, C., & Snyder, L., (1990). A comparison of programming models for shared memory multiprocessors. In: *ICPP* (2) (Vol. 1, pp. 163–170).
65. Luo, J. C., & Friedman, M. B., (1990). A parallel computational model for the finite element method on a memory-sharing multiprocessor computer. *Computer Methods in applied mechanics and Engineering, 84*(2), 193–209.
66. Maeda, R. K., Yang, P., Wu, X., Wang, Z., Xu, J., Wang, Z., & Wang, Z., (2016). JADE: A heterogeneous multiprocessor system simulation platform using recorded and statistical application models. In: *Proceedings of the 1st International Workshop on Advanced Interconnect Solutions and Technologies for Emerging Computing Systems* (Vol. 1, pp. 1–6).

67. Marsan, M. A., & Gerla, M., (1982). Markov models for multiple bus multiprocessor systems. *IEEE Transactions on Computers, 31*(03), 239–248.

68. Marszalek, W., & Podhaisky, H., (2016). 2D bifurcations and Newtonian properties of memristive Chua's circuits. *EPL (Europhysics Letters), 113*(1), 10005.

69. Martin, M. M., Sorin, D. J., Beckmann, B. M., Marty, M. R., Xu, M., Alameldeen, A. R., & Wood, D. A., (2005). Multifacet's general execution-driven multiprocessor simulator (GEMS) toolset. *ACM SIGARCH Computer Architecture News, 33*(4), 92–99.

70. McLeod, P., (1977). A dual task response modality effect: Support for multiprocessor models of attention. *Quarterly Journal of Experimental Psychology, 29*(4), 651–667.

71. Meier, H. M., Döscher, R., & Faxén, T., (2003). A multiprocessor coupled ice-ocean model for the Baltic sea: Application to salt inflow. *Journal of Geophysical Research: Oceans, 108*(C8), (Vol. 1, pp. 2–7).

72. Mishra, B. S. P., & Dehuri, S., (2011). Parallel computing environments: A review. *IETE Technical Review, 28*(3), 240–247.

73. Mishra, S., Schwab, C., & Šukys, J., (2012). Multi-level monte Carlo finite volume methods for nonlinear systems of conservation laws in multi-dimensions. *Journal of Computational Physics, 231*(8), 3365–3388.

74. Moller, F., & Tofts, C., (1990). A temporal calculus of communicating systems. In: *International Conference on Concurrency Theory* (Vol. 1, pp. 401–415). Springer, Berlin, Heidelberg.

75. Moore, C., (1997). Majority-vote cellular automata, ising dynamics, and P-completeness. *Journal of Statistical Physics, 88*(3), 795–805.

76. Mukherjee, R., & Bennett, J., (1990). Simulation of parallel computer systems on a shared memory multiprocessor. In: *Proc. of the 23rd Hawaii Int. Conf. on System Sciences* (Vol. 1, pp. 2–5).

77. Nishiura, D., & Sakaguchi, H., (2011). Parallel-vector algorithms for particle simulations on shared-memory multiprocessors. *Journal of Computational Physics, 230*(5), 1923–1938.

78. Orlov, S. P., & Efimushkina, N. V., (2016). Simulation models for parallel computing structures. In: *2016 XIX IEEE International Conference on Soft Computing and Measurements (SCM)* (Vol. 1, pp. 231–234). IEEE.

79. Padua, D., (2011). *Encyclopedia of Parallel Computing* (Vol. 1, pp. 2–4). Springer Science & Business Media.
80. Park, H. W., Oh, H., & Ha, S., (2009). Multiprocessor SoC design methods and tools. *IEEE Signal Processing Magazine, 26*(6), 72–79.
81. Pashler, H., (1990). Do response modality effects support multiprocessor models of divided attention?. *Journal of Experimental Psychology: Human Perception and Performance, 16*(4), 826.
82. Paulin, P. G., Pilkington, C., Langevin, M., Bensoudane, E., Lyonnard, D., Benny, O., & Nicolescu, G., (2006). Parallel programming models for a multiprocessor SoC platform applied to networking and multimedia. *IEEE Transactions on Very Large Scale Integration (VLSI) Systems, 14*(7), 667–680.
83. Peir, J. K., (1986). *Program Partitioning and Synchronization on Multiprocessor Systems (Parallel, Computer Architecture, Compiler)* (Vol. 1, pp. 2–9). University of Illinois at Urbana-Champaign.
84. Poletti, F., Poggiali, A., Bertozzi, D., Benini, L., Marchal, P., Loghi, M., & Poncino, M., (2007). Energy-efficient multiprocessor systems-on-chip for embedded computing: Exploring programming models and their architectural support. *IEEE Transactions on Computers, 56*(5), 606–621.
85. Poplavko, P., Basten, T., Bekooij, M., Van, M. J., & Mesman, B., (2003). Task-level timing models for guaranteed performance in multiprocessor networks-on-chip. In: *Proceedings of the 2003 International Conference on Compilers, Architecture and Synthesis for Embedded Systems* (Vol. 1, pp. 63–72).
86. Ramamritham, K., Stankovic, J. A., & Shiah, P. F., (1990). Efficient scheduling algorithms for real-time multiprocessor systems. *IEEE Transactions on Parallel and Distributed Systems, 1*(2), 184–194.
87. Revol, N., & Théveny, P., (2014). Numcrical reproducibility and parallel computations: Issues for interval algorithms. *IEEE Transactions on Computers, 63*(8), 1915–1924.
88. Rosing, M., Schnabel, R. B., & Weaver, R. P., (1992). Scientific programming languages for distributed memory multiprocessors: Paradigms and research issues. *Advances in Parallel Computing, 3*, 17–37.
89. Sadegh, N., (1993). A perceptron network for functional identification and control of nonlinear systems. *IEEE Transactions on Neural Networks, 4*(6), 982–988.

90. Schliecker, S., Rox, J., Ivers, M., & Ernst, R., (2008). Providing accurate event models for the analysis of heterogeneous multiprocessor systems. In: *Proceedings of the 6th IEEE/ACM/IFIP International Conference on Hardware/Software Codesign and System Synthesis* (Vol. 1, pp. 185–190).
91. Sengupta, A., & Dahbura, A. T., (1992). On self-diagnosable multiprocessor systems: Diagnosis by the comparison approach. *IEEE Transactions on Computers, 41*(11), 1386–1396.
92. Sharma, G., & Martin, J., (2009). MATLAB®: A language for parallel computing. *International Journal of Parallel Programming, 37*(1), 3–36.
93. Skillicorn, D. B., (1990). Architecture-independent parallel computation. *Computer, 23*(12), 38–50.
94. Smith, D. R., (1993). Derivation of parallel sorting algorithms. In: *Parallel Algorithm Derivation and Program Transformation* (Vol. 1, pp. 55–69). Springer, Boston, MA.
95. Sokolinsky, L. B., (2021). BSF: A parallel computation model for scalability estimation of iterative numerical algorithms on cluster computing systems. *Journal of Parallel and Distributed Computing, 149*, 193–206.
96. Sorel, Y., (1994). Massively parallel computing systems with real time constraints: The "algorithm architecture adequation" methodology. In: *Proceedings of the First International Conference on Massively Parallel Computing Systems (MPCS) The Challenges of General-Purpose and Special-Purpose Computing* (Vol. 1, pp. 44–53). IEEE.
97. Squillante, M. S., & Lazowska, E. D., (1993). Using processor-cache affinity information in shared-memory multiprocessor scheduling. *IEEE Transactions on Parallel and Distributed Systems, 4*(2), 131–143.
98. Sunderam, V. S., (1990). PVM: A framework for parallel distributed computing. *Concurrency: Practice and Experience, 2*(4), 315–339.
99. Tosic, P. T., (2004). A perspective on the future of massively parallel computing: Fine-grain vs. coarse-grain parallel models comparison & contrast. In: *Proceedings of the 1st Conference on Computing Frontiers* (Vol. 1, pp. 488–502).
100. Tsuei, T. F., & Yamamoto, W., (2003). Queuing simulation model for multiprocessor systems. *Computer, 36*(2), 58–64.

101. Vella, K., (1998). *Seamless Parallel Computing on Heterogeneous networks of Multiprocessor Workstations* (Vol. 1, pp. 2–9). Doctoral dissertation, University of Kent at Canterbury.

102. Veltman, B., Lageweg, B. J., & Lenstra, J. K., (1990). Multiprocessor scheduling with communication delays. *Parallel Computing, 16*(2, 3), 173–182.

103. Waintraub, M., Schirru, R., & Pereira, C. M., (2009). Multiprocessor modeling of parallel Particle Swarm Optimization applied to nuclear engineering problems. *Progress in Nuclear Energy, 51*(6, 7), 680–688.

104. Wang, J. J., & Xiao, A. G., (2018). An efficient conservative difference scheme for fractional Klein–Gordon–Schrödinger equations. *Applied Mathematics and Computation, 320*, 691–709.

105. Wilkinson, P., (2006). *Parallel Programming: Techniques and Applications Using Networked Workstations and Parallel Computers* (2nd edn., Vol. 1, pp. 2, 3). Pearson Education India.

106. Wolf, W., Jerraya, A. A., & Martin, G., (2008). Multiprocessor system-on-chip (MPSoC) technology. *IEEE Transactions on Computer-Aided Design of Integrated Circuits and Systems, 27*(10), 1701–1713.

107. Wu, C. H., Hodges, R. E., & Wang, C. J., (1991). Parallelizing the self-organizing feature map on multiprocessor systems. *Parallel Computing, 17*(6, 7), 821–832.

108. Zeng, Z. Q., Yu, H. B., Xu, H. R., Xie, Y. Q., & Gao, J., (2008). Fast training support vector machines using parallel sequential minimal optimization. In: *2008 3rd International Conference on Intelligent System and Knowledge Engineering* (Vol. 1, pp. 997–1001). IEEE.

109. Zhang, K., Zhang, H. G., Cai, Y., & Su, R., (2019). Parallel optimal tracking control schemes for mode-dependent control of coupled Markov jump systems via integral RL method. *IEEE Transactions on Automation Science and Engineering, 17*(3), 1332–1342.

CHAPTER 5

DISTRIBUTED COMPUTING

CONTENTS

5.1. INTRODUCTION

The distributed system is a collection of autonomous entities that work together to solve an issue that cannot be solved separately. Distributed systems have usually been around since the beginning of time. There exists communication between mobile sentient agents in nature or between a group of fish to the group of birds to complete ecosystems of microbes. The concept of distributed systems as a valuable and extensively deployed technology is now becoming a reality, thanks to the broad utilization of the Internet and the emergence of the globalized world (Figure 5.1).

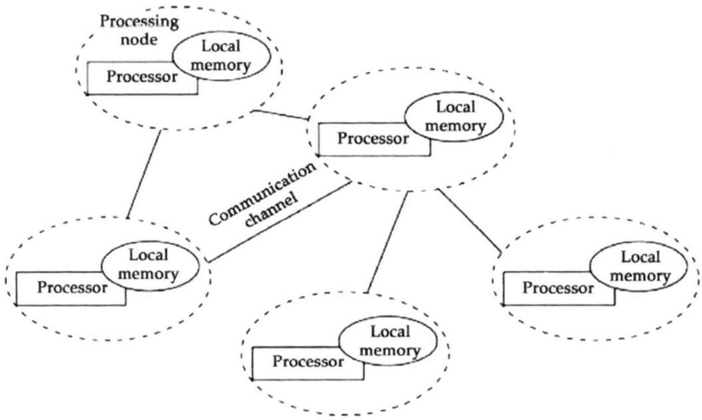

Figure 5.1. Distributed computing system.

Source: https://www.sciencedirect.com/topics/computer-science/distributed-computing.

A distributed system can be described in several ways in computing systems:

- You know you are using one when the crash of a computer you have never heard of prevents you from doing work (Alboaie).
- The group of computers that do not share memory or a physical clock, connect by messages traveling across a communication network and every computer seems to have its specific memory and operating system. Generally, the computers are semi-autonomous and loosely linked when they work together to solve a problem (Singhal and Shivaratri, 1994).
- A system that comprises a set of individual computers that seem to users as a single cohesive computer (Van Steen and Tanenbaum, 2002).

- A word that encompasses a broad variety of computer systems, varying from loosely coupled systems like wide-area networks (WANs) to firmly coupled systems like local area networks (LANs) to extremely strongly coupled systems like multiprocessor systems (Goscinski, 1991).

The distributed system is a group of largely autonomous processors interacting across the communication system and exhibiting the following characteristics:

- **No Shared Physical Clock:** This is a crucial assumption since it introduces the concept of "distribution" into the system and leads to the processors' intrinsic asynchrony.
- **No Common Memory:** It is a critical characteristic that necessitates message transfer for communication. This characteristic involves the elimination of a shared physical clock. It should be emphasized that the distributed system can still provide the assumption of a shared address space through the distributed common memory abstraction. Numerous characteristics of common memory multiprocessor systems have been examined in the research on distributed computing.
- **Geographical Separation:** The further separated the processors are physically, the more representational the architecture of the distributed system is. However, the processors are not required to be connected to WAN. The web of workstations (COW/NOW) setup integrating processors on the LAN is increasingly viewed as a tiny distributed system as of late. This NOW arrangement is gaining popularity due to the widely accessible cheap, high-speed processors. Google's search engine employs the NOW design.
- **Autonomy and Heterogeneity:** The processors are loosely connected, meaning they can run different operating systems and have varied speeds. They are not usually part of the specific system, but they do work together to provide services or solve problems.

5.2. ASSOCIATION TO COMPUTER SYSTEM MODULES

Figure 5.2 depicts a usual distributed system. Every computer possesses unit for processing of memory, and the computers are linked via a network. Figure 5.3 depicts the associations between the software modules that run on every computer and rely on operating system and the network protocol for operation. Middleware is another term for distributed software. Distributed implementation is the implementation of procedures across a distributed system to accomplish a shared goal collectively. An execution is often known as a run or a computation.

The distributed system has a layered approach to reduce the design complexity. Middleware is the distributed program that controls the distributed system and provides heterogeneous platform-level transparency (Lea, Vinoski, and Vogels, 2006).

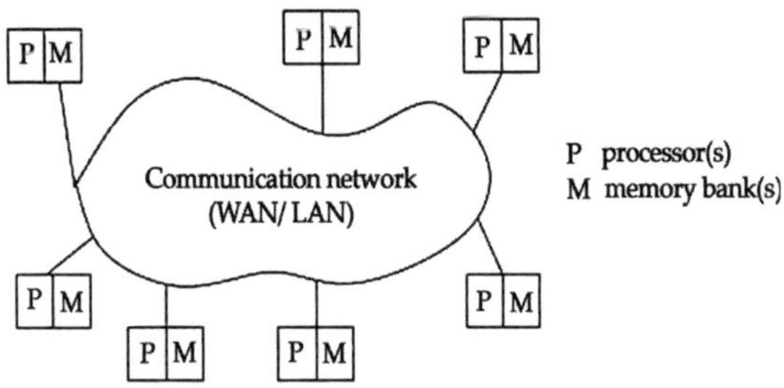

Figure 5.2. A communication system links processors in a distributed system.

Source: https://www.amazon.com/Distributed-Computing-Principles-Algorithms-System/dp/05211898455.

The collaboration between these system modules at every processor is depicted in Figure 5.3. The typical application layer operations of the network protocol, like http, mail, and telnet, are not included in the middleware layer. The user application code contains primitives and invites to functions specified in the middleware layer's different libraries.

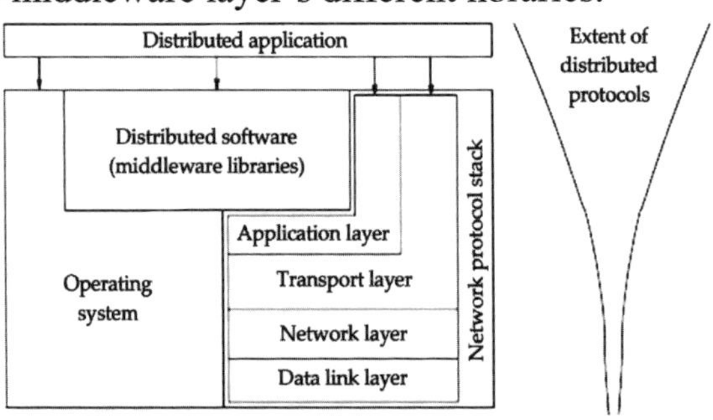

Figure 5.3. Communication of the software modules at every processor.

Source: https://www.amazon.com/Distributed-Computing-Principles-Algorithms-System/dp/05211898455.

There are numerous libraries from which to pick when invoking primitives for the typical middleware layer functions, like dependable and ordered multicasting. There are numerous standards, like the CORBA (common object request broker architecture) of the OMG (Object Management Group) (Vinoski, 1997) and the RPC (remote procedure call) mechanism (Ananda, Tay, and Koh, 1992; Coulouris, Dollimore, and Kindberg, 2005). The RPC method functions conceptually like just a local procedure call, with the exception that the method code might reside on the remote machine and the RPC program transmits a network message to execute the remote procedure. After waiting for a response, the procedure call is concluded from the viewpoint of the calling software. Commercially available versions of middleware frequently employ CORBA, DCOM, and RMI (remote method invocation) techniques (Campbell, Coulson, and Kounavis, 1999). MPI (message-passing interface) (Gropp and Lusk, 1994; Snir et al., 1998) is an instance of an interface for numerous communication operations.

5.3. MOTIVATION

All or most of the following conditions motivate the use of a distributed system:

- **Inherently Distributed Computations:** The computing is intrinsically distributed in several applications, like transfer of funds in banking or attaining consensus between geographically distant participants.
- **Sharing of Resource:** Resources like peripherals, entire data tables in databases, specific libraries, and data cannot be fully reproduced at all sites since it is not possible nor economical. In addition, they cannot be located in a single location since access to that location could become congested. Consequently, such resources are often dispersed throughout the system. Distributed databases, like DB2, partition data sets across multiple servers further to duplicating them at some sites for quick access and high availability.

Access to data and resources that are located in different parts of the world. The data cannot be duplicated at each site collaborating in the distributed execution in several scenarios since it is either too huge or too sensitive. Payroll data, for instance, is too huge and too delicate to be copied at each branch office of a multinational firm. As a result, it is saved on a central server that can be accessed by branch offices. Likewise, exceptional resources like supercomputers are only available in specific places, and users must check in remotely to utilize them.

Improvements in the engineering of resource-reserved mobile devices, along with developments in the wireless technologies used to connect with these devices, have increased the significance of distributed protocols and middleware.

- **Improved Dependability:** Due to the capability of duplicating resources and executions, along with the fact that geographically dispersed resources are not likely to fail simultaneously under normal conditions, a distributed method has the capacity to provide greater reliability. Reliability involves numerous aspects:
- **Availability:** for example, the resource must always be available.
- **Integrity:** for example, the value of the resource must be accurate despite concurrent access from several processors, in accordance with the semantics anticipated by the application.
- **Fault-Tolerance:** For example, the capacity to recoup from system problems, where such breakdowns might well be described as occurring according to one of several failure models.
- **Enhanced Cost-to-Performance Ratio:** This ratio is improved via resource sharing and retrieving geographically distant data and resources. Even though increased throughput is not always the primary goal of adopting a distributed system, each task can be segmented over the multiple processors in a distributed system. When compared to employing customized parallel computers, this architecture delivers a higher performance/cost ratio. This is especially true in the case of the NOW setup.

A distributed system has the following benefits further to achieving the above requirements:

- **Scalability:** As processors are typically interconnected via a WAN, the addition of more processors does not directly impede the communications system.
- **Modularity and Incremental Expandability:** Until all processors are executing the same middleware processes, heterogeneous processors can be added to the system without negatively impacting its performance. Likewise, it is simple to replace old processors with other processors.

5.4. ASSOCIATION TO PARALLEL MULTIPROCESSOR/MULTICOMPUTER SYSTEMS

Above, the features of the distributed system were pointed out. Figure 5.2 depicts an example of a distributed system. How, on the other hand, does a person categorize a system that fits many but not all the criteria? Is the system still distributed or has it evolved into the parallel multiprocessor system? To properly find the answers, the design of parallel systems will be examined first, followed by some famous classifications for multiprocessor systems.

5.4.1. Features of Parallel Systems

This system might be widely categorized as fitting to among these three kinds. Several processors have explicit access to common memory, which constitutes a shared address space, in the multiprocessor system, which is typically a parallel system. The structure is depicted in Figure 5.4. Such processors typically lack a shared clock.

A multiprocessor system is frequently associated with UMA (uniform memory access) topology, in which the accessing latency, or the time it takes for each processor to perform an access to every memory address, is the similar. The processors are physically close together and are linked via an interconnection network. Read and write functions on shared memory have traditionally been used to communicate between processors, though message-passing components like those offered by the MPI can also be used. The operating system that runs on all of the processors is typically the same, and the software and hardware are tightly connected.

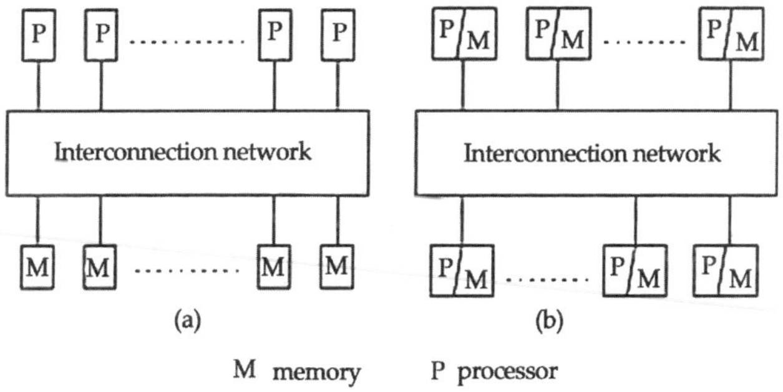

Figure 5.4. Two typical designs for parallel systems. (a) UMA multiprocessor system; (b) NUMA (non-uniform memory access) multiprocessor. In both of the architectures, the processors might cache data locally from memory.

Source: https://www.amazon.com/Distributed-Computing-Principles-Algorithms-System/dp/05211898455.

Processors of the similar type are typically placed in almost the same container with shared memory. The memory interconnection network might be a bus, but for improved competence, it is commonly a multiple stage switch having regular and symmetric architecture.

Figure 5.5 depicts two prominent networks for interconnection: the Omega (Beneš, 1965) and the Butterfly (Cooley and Tukey, 1965), both of which are multi-stage networks with 2 switching elements. Data from any of the 2 input wires can be shifted to the lower or upper output wire with every 2 × 2 switch. Just one unit of data can be delivered on the output wire in the single step. There is a clash if data from all of the input wires is sent to the same output wire in the single step. Collisions can be avoided using a variety of strategies like buffering or more complex connectivity architectures.

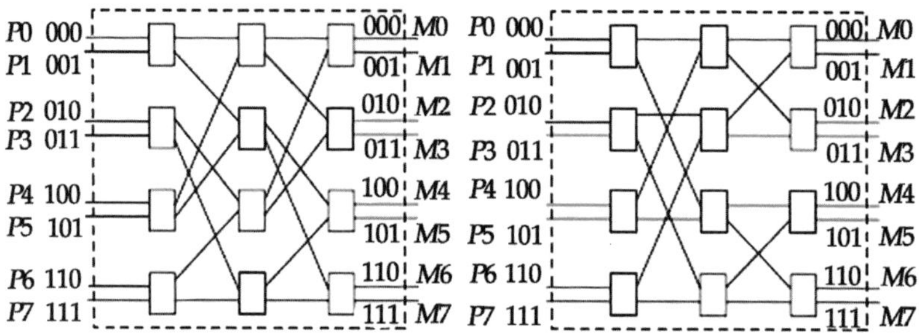

Figure 5.5. Networks of interconnection for common memory multiprocessor systems. (a) Omega network for n = eight processors P_0, P_1, …, P_7 and memory banks M_0, M_1, …, M_7. (b) Butterfly network for n = eight processors P_0, P_1, …, P_7 and memory banks M_0, M_1, …, M_7.

Source: https://eclass.uoa.grr/modules/document/file.php/D2245/2015/DistrComp.pdf.

In the figure, each 2 × 2 switch is depicted as a rectangle. In addition, a network with n inputs and n outputs utilizes log n bits and log n stages for routing. Routing in 2 × 2 switch at stage k employs only the kth bit and may therefore be performed in hardware at speed of clock. Multistage networks can generally be formed recursively, and the connectivity pattern among any 2 stages can be described using either a recursive or iterative generating function. In addition to the Butterfly and Omega networks, the Clos (1953) and shuffle-exchange networks (Wu and Feng, 1980) are instances of multistage interconnection networks. Every one of them possesses extremely intriguing mathematical features that permit for extensive communication amongst the memory and processor bank.

5.4.1.1. Omega Interconnection Function

n/2log2 n switching components of size 2 × 2 are placed in log2 n stages in the Omega network, which links *n* processors to *n* units *o* memory. A link prevails amongst the output i of 1 stage and the input j of another stage within every pair of neighboring stages of Omega network, as per preceding ideal shuffle sequence, which is the left-rotation function on the binary sequence of i to get j. The following is the repetitive generating function.

Consider all switch stages. Casually, the upper input lines for every switch originate from upper half of the switches in the previous stage in sequential manner.

n = 8 in the Omega network shown in Figure 5.5(a). As a result, the output i of some stage is linked to the next stage's input 2i for outputs i where $0 \leq i \leq 3$. For $4 \leq i \leq 7$, any stage's output i is coupled to the next stage's input 2i + 1 – n.

$$j = \begin{cases} 2i, & \text{for } 0 \leq i \leq n/2 - 1, \\ 2i + 1 - n, & \text{for } n/2 \leq i \leq n - 1. \end{cases}$$

5.4.1.2. Omega Routing Function

Just j and the stage number s are considered in this operation from an input line i to an output line j, where s is [0, log2n 1]. In the stage s switch, data is transmitted to the higher output wire if the s + 1st MSB of j is zero, else it is directed to the lower output wire. Due to the obvious routing algorithm described below, it makes no difference if the 2 incoming connections travel to the lower or upper input port.

5.4.1.3. Butterfly Routing Function

In the stage s switch, data is sent to the upper output wire if s + 1st MSB of j is zero, else it is directed to the lower wire.

Notice that the pathways from the various inputs to any output in the Omega and Butterfly networks create a spanning tree. When the data is bound for the very same output line, encounters are inevitable. The benefit, however, is that data can be merged at the switches if the program interpretations are recognized.

5.4.1.4. Multicomputer Parallel Systems

In this type of system, the multiple processors *do not have full access to common memory.* Multiple processors' memory might or might not constitute a shared address space. A common clock is generally absent from such computers. Figure 5.4 depicts the design.

The processors are located in close vicinity, are typically tightly integrated, and are linked through the interconnection network. The processors interact through a shared address space or by exchanging messages. A multicomputer system with a common address space typically correlates to a NUMA design in which the delay to access multiple common memory locations from the various processors differs. The wrap-around 4 × 4 mesh is shown in Figure 5.6. The greatest path length among any 2 processors in a k × k mesh containing k processors is 2(k/2 1). The Manhattan grid can generally be used for routing. A 4D hypercube is depicted in Figure 5.6. 2k processor and memory units make up a k-dimensional hypercube (Feng, 1981; Harary, Hayes, and Wu, 1988). Each of these units is basically a node in this hypercube with its own k-bit label. Every one of the k dimensions in the descriptor corresponds to a bit position. Excluding the position of bit relating to the dimension in which the 2 nodes vary, the titles of any 2 adjacent nodes are similar. As a result, the processor labels are labeled in such a way that the quickest route among any 2 processors is typically the Hamming distance (described as the quantity of bit positions in which 2 equal sized bit strings diverge). This is clearly defined by the value of k.

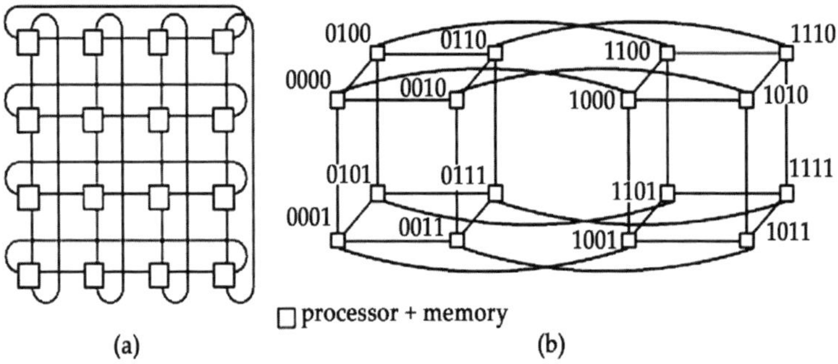

Figure 5.6. Some common configurations for multicomputer common-memory computers. (a) Wrap-around two-dimensional-mesh, also called torus; (b) hypercube with dimension 4.

Source: https://eclass.uoa.grr/modules/document/file.php/D2245/2015/DistrComp.pdf.

Inside the hypercube, configuration is done hop by hop. The message can be transmitted along every dimension relating to the bit position where the existing node's address differs from the destination node at whichever hop. The figure shows a four-dimensional hypercube constituted by linking the respective edges of 2 three-dimensional (3D) hypercubes (correlating to the right and left "cubes" in figure) all along 4th dimension; the identifiers of the four-dimensional hypercube are created by associating a "0" to the identifiers of the left 3D hypercube and associating a "1" to the identifiers of the right 3D hypercube. This can be expanded to create higher-dimensional hypercubes. Take note of the fact that there are numerous paths among any two nodes, providing fault tolerance along with the blocking control method. The hypercube and its derivative topologies have very fascinating mathematical features with routing and fault-tolerance consequences.

Array processors are a type of parallel computer that is physically co-located, very strongly connected, and shares a system clock (but might not share memory and interact by passing data with the help of messages). This category includes systolic arrays and array processors that execute closely coordinated processing and data interchange in lock-step for purposes like DSP and processing of image. These applications typically entail a significant amount of data repetitions. This type of parallel system has a small market.

The gap amongst NUMA multiprocessors and UMA and message-passing multicomputers is significant since design of algorithm and data and job division between processors must accommodate for uncertain and volatile latencies in communication/memory (Grama, Kumar, Gupta, and Karypis, 2003). NUMA and message-passing systems, in comparison to array processors and UMA, are quite less appropriate when the level of granularity of acquiring data access and communication is indeed very high.

5.4.2. Flynn's Taxonomy

Flynn (1972) defined 4 processing modes centered on whether the processors implement the same or separate instruction streams simultaneously and whether or not they analyze the same data simultaneously (Figure 5.7).

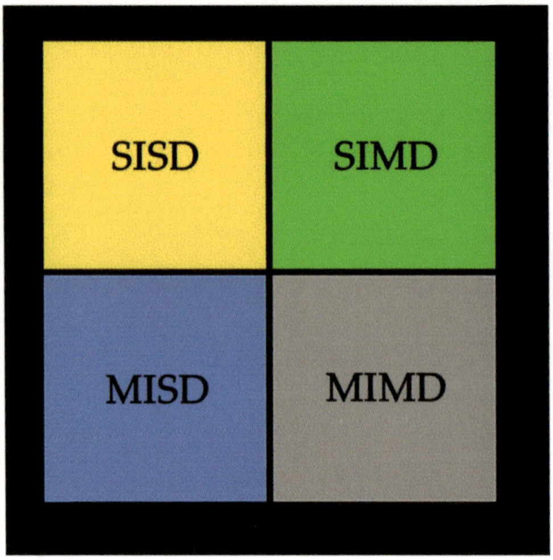

Figure 5.7. Flynn's taxonomy categorizes how a processor implements instructions.

Source: https://diveintosystems.org/book/C15-Parallel/index.html.

This classification is informative for understanding the variety of system configuration options.

5.4.2.1. SISD (Single Instruction Stream, Single Data Stream)

This mode is equivalent to the von Neumann model's traditional processing with a single central processing unit (CPU) and the single memory unit coupled by the system bus.

5.4.2.2. SIMD (Single Instruction Stream, Multiple Data Stream)

This mode resembles to the execution of many homogeneous processes on separate data items in lock-step. Applications that entail operations on huge matrices and arrays, like scientific applications, benefit the most from computers that support SIMD operation mode since data sets can be conveniently partitioned.

SIMD machines comprised a number of the initial parallel computers, including Illiac-IV, CM2, MasPar MP-1, and MPP. Vector processors, systolic arrays, and array processors are also included in the SIMD processing category. Modern SIMD designs consist of co-processing modules like the MMX units in Intel CPUs (for example, Pentium with SSE (streaming SIMD extensions) capabilities) and DSP chips like the Sharc (Grama et al., 2003).

5.4.2.3. MISD (Multiple Instruction Stream, Single Data Stream)

This mode resembles to the simultaneous execution of many operations on the same data. This is a highly specialized operation mode having a small number of yet specific applications, such as visualization.

5.4.2.4. MIMD (Multiple Instruction Stream, Multiple Data Stream)

In this method, each CPU runs distinct code on diverse information. This is the manner of operation in both distributed systems and the almost all of parallel systems. The system processors do not share a clock. Machines that run in MIMD mode include Sun Ultra servers, IBM SP machines, and multicomputer PCs.

In Figure 5.8, the MISD, SIMD, and MIMD designs are depicted. MIMD designs are the most common and offer the most versatility in dividing code and data between processors. MIMD designs also incorporate the conventionally understood manner of distributed system implementation.

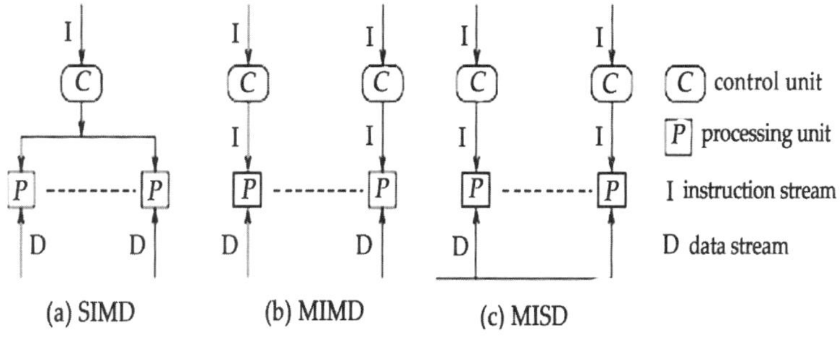

Figure 5.8. Flynn's taxonomy of MIMD, SIMD, and MISD designs for multicomputer/multiprocessor systems.

Source: https://eclass.uoa.gr/modules/document/file.php/D245/2015/DistrComp.pdf.

5.4.3. Coupling, Concurrency, Parallelism, and Granularity

- **Coupling:** The extent of coupling between a collection of components, whether software or hardware, is measured by their interdependency, binding, or homogeneity. The components are considered to be tightly connected when the extent of coupling is high. The mutual timing of the shared data or the shared instruction stream causes MISD and SIMD designs to be strongly connected. In this section, several MIMD designs in terms of coupling are looked at:

- Tightly connected multiprocessors (having UMA common memory). These might be either based on switch (for example, NYU Ultracomputer) or based on bus (for example, Sequent).
- Multiprocessors with close coupling (having NUMA common memory or that interact by message passing). SGI Origin 2000 and Sun Ultra HPC servers (which interact through NUMA common memory), along with the torus and hypercube (that interact by message passing), are examples.
- Co-located physically loosely connected multicomputers (without common memory). These can be based on bus (for example, NOW linked by a Myrinet card or LAN) or use a more typical communication network, with heterogeneous CPUs. Processors in these kinds of systems do not share common clock or memory, and so might be categorized as distributed networks — yet, the processors are quite close to each other, indicating a parallel system. Because communication latency in these kinds of systems might be much shorter as compared to wide-area distributed systems, solution strategies to multiple issues might differ from those in wide area distributed systems.
- Physically distant loosely connected multicomputers (without common memory or the common clock). These resemble to the traditional concept of distributed structures.

- **Parallelism or Acceleration of a Program on a Particular System:** This is a measurement of a program's relative acceleration on a specific machine. The speedup is determined by the amount of processors and the code's mapping to the processors. It is calculated as the proportion of time T(1) for a single processor to time T(n) for n processors.
- **Parallelism within the Distributed Program:** This is an overall measure of the proportion of time that almost all processors are

actively implementing CPU instructions rather than waiting for communication (through message-passing or common memory) processes to complete. The phrase has typically been utilized to describe parallel programs. If the overall measure is just an operation of the code, then parallelism is unaffected by the design. Eventually, this term becomes synonymous with parallelism.

- **Concurrency of the Program:** This is a wider term with the same general meaning as parallelism of the program, but it is utilized in the perspective of distributed applications. The parallelism of a distributed program is characterized as the proportion of local (non-shared access of memory and non-communication) functions to the entire amount of operations, which includes communication or common memory access operations.
- **Granularity of the Program:** Granularity is the proportion of the amount of processing to the amount of communication in a distributed application. When the extent of parallelism is fine-grained, there are fewer productive CPU instruction operations relative to the amount of times the processors connect with one another through message-passing or common memory and need to be synchronized. Tightly connected systems are best served by programs having fine-grained parallelism. ISD and SIMD designs, closely connected MIMD multiprocessors (with common memory), and weakly coupled multiprocessors (without common memory) that are co-located physically are all examples of this. The latency glitches for regular communication across the WAN would severely decrease the overall performance if fine-grained parallel applications were executed over loosely connected multiprocessors that were relatively far. As a result, programs having the coarse-grained messaging granularity will suffer significantly less overhead on these loosely connected multicomputers. The partners amongst the local functioning machine, the middleware that implements the distributed software, and the network protocol are shown in Figure 5.3.

5.5. MESSAGE-PASSING SYSTEMS VS. SHARED MEMORY SYSTEMS

Shared memory systems have a common address space that runs all across the system. Communication between processors occurs through control and shared data variables for synchronization. Semaphores, which were originally created for common memory multiprocessors and uniprocessors, are instances of how shared memory systems might accomplish synchronization. Message passing is used by all multicomputer systems that lack a common address space given by the underlying design and hardware. In theory, programmers find common memory simpler to use as compared to message forwarding. For this and various other reasons, which will be discussed later, the concept known as common memory is often used to simulate a common address space. This concept is known as distributed common memory in the distributed system. Designing this concept involves a cost, but it enhances the application programer's job. A famous folklore conclusion is that message passing communication can be simulated by common memory communication and vice versa. As a result, the 2 paradigms are interchangeable (Figure 5.9).

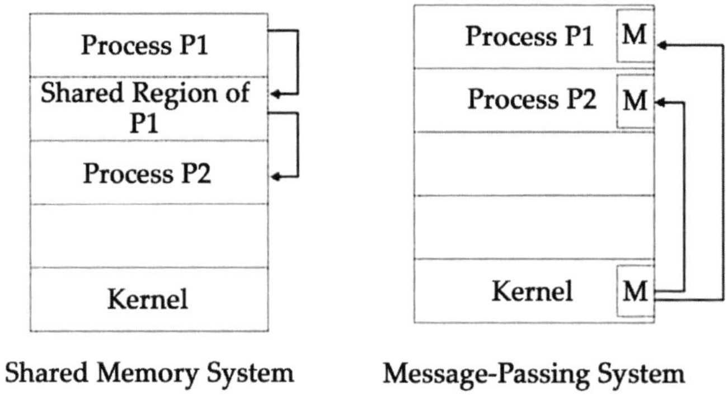

Figure 5.9. The message-passing model and the shared memory model.

Source: https://beingintelligent.com/difference-between-shared-memory-and-message-passing-process-communication.html.

5.5.1. Mirroring Message-Passing on the Shared Memory System

The common address space can generally be divided into several portions, with each processor receiving its own portion. Receive and send operations can be done by reading and writing from the address space of the sender/destination processor, correspondingly. Particularly, a mailbox can be allocated for every pair of processes in a different location. A P_i–P_j message exchange can be mimicked by P_i writing to the mailbox followed by P_j

reading from the mailbox. In the most basic scenario, the size of these kind of mailboxes is supposed to be limitless. Controlling the read and write operations with synchronization primitives in order to alert the sender/receiver when the data has been received/sent.

5.5.2. Mirroring Shared Memory on the Message-Passing System

For read and write operations, the receive and send functions are used. A write operation to the common position is mimicked by transmitting an update message to the relevant owner procedure; a read to the common location is mirrored by sending the query message to the owner procedure. Mirroring is costly since accessing some other processor's memory necessitates receive and send operations. Even though it may appear that mimicking common memory is more appealing from a programer's standpoint, it should be noted that it is merely a concept in the distributed system. Since write and read operations are executed via network-wide communication behind the covers, latency in write and read operations might be substantial even when utilizing shared memory mirroring.

Of course, an application can use an amalgamation of message-passing and shared memory. Every processor in the MIMD message passing multicomputer system might be a tightly connected multiprocessor system with common memory. The processors in the multiprocessor system interact with one another through shared memory. Message forwarding is used to communicate among two computers. Because message-passing systems are popular and better adapted for wide-area distributed systems, they will be looked at more thoroughly as compared to the common memory systems.

5.6. PRIMITIVES FOR DISTRIBUTED COMMUNICATION

5.6.1. Non-Blocking/Blocking, Asynchronous/Synchronous Primitives

Receive() and Send() are communication primitives for receiving and sending messages, respectively. A Send primitive requires two parameters as a minimum: the destination and the user-space buffer comprising the information to be transmitted. Likewise, a Receive primitive requires two parameters as a minimum: the source from which information is to be received (which might be a wildcard) and the user buffer into which information is to be received.

When using the send primitive, one might choose between delivering data in an unbuffered or buffered manner. The information is copied from a user buffer to the kernel buffer using the buffered option. The data is then transmitted to the network from a kernel buffer. The information is transferred straight from the customer buffer over onto network with the unbuffered option. The buffered option is normally necessary for the receive primitive since the data might have reached when the primitive is activated and requires a storage location in the kernel (Figure 5.10).

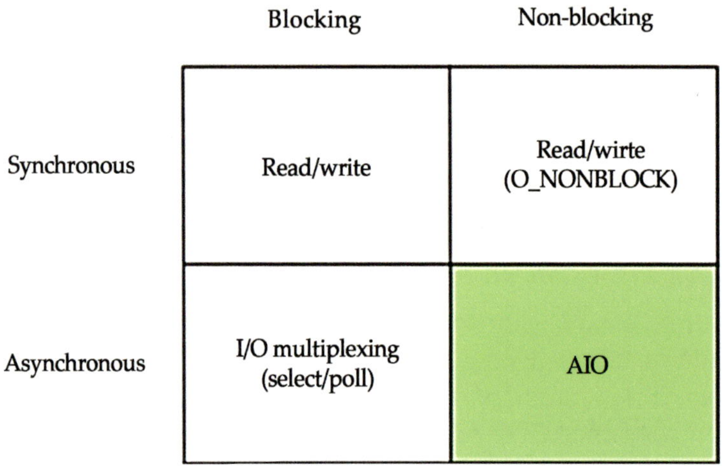

Figure 5.10. Non-blocking/blocking, asynchronous/synchronous primitives.

Source: https://sdesigner.tistory.com/88.

The terms non-blocking/blocking and asynchronous/synchronous primitives are defined in the following sections (Cypher and Leu, 1994):

- **Synchronous Primitives:** If the Receive() and Send() and primitives handshake with one another, the primitive is synchronous. Only until the initiating processor knows that the added equivalent receive primitive has been triggered and that the receive function has been finished does the execution for the send primitive complete. When the information to be obtained is copied into the receiver's user buffer, the receive primitive's processing is complete.

- **Asynchronous Primitives:** When control comes back to the calling process after the item of data to be transmitted has been transferred out of the user-specified buffer, the send primitive is called asynchronous.

It doesn't make any sense to describe asynchronous receive primitives.

- **Blocking Primitives:** If control comes back to the calling process after the primitive's processing (whether in asynchronous or synchronous mode) is complete, the primitive is blocking.
- **Non-Blocking Primitives:** The primitive is non-blocking if the control is instantly returned to the calling process upon initiation, even if the action hasn't yet finished. For the non-blocking send, control goes to the process prior to copying the data from the operator buffer. For the non-blocking receive, control is returned to the process before the information from the sender has arrived.

A return variable on the non-blocking primitive call reverts a system-produced handle that can be utilized to validate the status of the call's completion later. In two manners, the process can verify that the call has been completed. First of all, it can validate (in loop or on the regular basis) whether the handle has indeed been flagged or uploaded. Secondly, the Wait set of handles as variables can be issued. If one of the variable handles is posted, the Wait call normally blocks. After executing the primitive in the non-blocking manner, the process has likely completed whatever operations it can and now wants to know the condition of the call's completion, hence employing a blocking Wait() call is standard programming practice. As demonstrated in Figure 5.11, the code for the non-blocking Send might look like this.

```
Send(X, destination, handle_k)                        // handle_k is a return parameter
...
...
Wait(handle_1, handle_2, ..., handle_k, ..., handle_m)  // Wait always blocks
```

Figure 5.11. The send primitive is non-blocking. At least one of the Wait call's parameters is posted when it returns.

Source: https://www.cs.uic.edu/~ajayk/DCS-Book.

When at the time that Wait() is given, the processing for the primitive has concluded, the Wait returns instantly. The value of handle k can be used to determine when the primitive's processing is complete. If the implementation of the primitive hasn't completed, the Wait pauses and waits for the signal to awaken it up. When the execution for the primitive finalizes, the communication system software fits the value of handle k and awakens any process having a wait call stuck on this $handle_k$. This is referred to as posting the operation's completion.

As a result, the send primitive has the following variants:

- Synchronous non-blocking;
- Synchronous blocking;
- Asynchronous non-blocking;
- Asynchronous blocking.

There exists non-blocking synchronous and blocking synchronous variants of the receive primitive.

Figure 5.12 uses a timing diagram to show these variations of the primitives. Every process is represented by three-time lines: (a) the processing of the process; (b) the user buffer to/from which data is received/sent; and (c) the communication/Kernel subsystem.

- **Blocking Synchronous Send (Figure 5.12(a)):** The information is transferred from the user to kernel buffer before being transmitted over the network. After the data has been copied to the receiver's system buffer and the Receive call has already been sent, a confirmation back to the sender wraps up the Send operation by returning command to the process that initiated the Send.
- **Non-Blocking Synchronous Send (Figure 5.12(b)):** As long as the data copy from the user to kernel buffer is started, control goes to the initiating process. The handle of a place that the user executes can later verify for the accomplishment of the synchronous send function is likewise set as a variable in the non-blocking call. According to the mechanics described above, the location is provided once the receiver sends an affirmation (i). By validating the returning handle, the user procedure can regularly check for the execution of non-blocking synchronous Send, or it might perform the blocking Wait function on the returned handle (Figure 5.12(b)).

- **Blocking Asynchronous Send (Figure 5.12(c)):** The user function that uses Send is paused till the buffer of user is copied to the kernel buffer. (For an unbuffered choice, the user function that uses Send is halted till the user's buffer is forwarded to the network.)
- **Non-Blocking Asynchronous Send (Figure 5.12(d)):** The user function that executes Send is halted until the data transfer from the users to kernel buffer begins. (In the case of the unbuffered choice, the user function that executes Send is halted until the data transfer from the user's buffer to a network is started.) As early as this transfer is started, control returns to the user function, and the variable in non-blocking call is generally fixed with the handle of a position that the user function can verify later with the help of Wait function for the execution of the asynchronous Send function. When the information has been moved out of the buffer of user, the asynchronous Send wraps up. If the user desires to reutilize the buffer from which the information was transmitted, verifying for completion might be required.
- **Blocking Receive (Figure 5.12(a)):** Receive call is blocked until the requested data is received and is written to the user buffer supplied. Control is then passed back to the user function.
- **Non-Blocking Receive (Figure 5.12(b)):** The Receive call will result in the kernel registering the call and returning a handle to the position where the user function can validate for the execution of non-blocking Receive function. This address is written to the user-specified buffer by the kernel after the anticipated data has arrived. The user function can determine whether the non-blocking Receive action has completed by performing the Wait function on the delivered handle. (If the information has already reached at the time of the call, it will be waiting in a kernel buffer and must be transferred to the user buffer.)

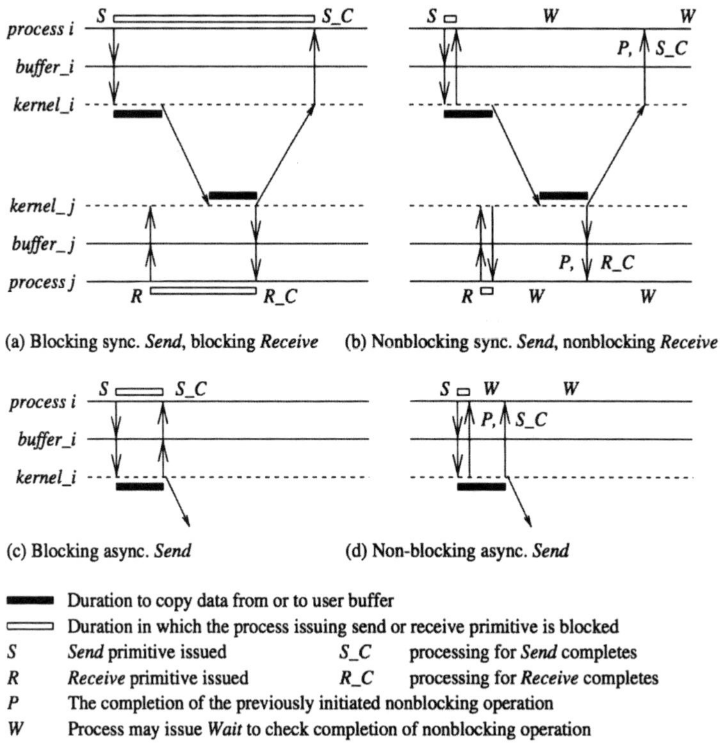

Figure 5.12. Non-blocking/blocking and asynchronous/synchronous primitives (Cypher and Leu, 1994). Process P_i is sending and P_j is receiving. (a) Blocking synchronous send and receive; (b) non-blocking synchronous send and receive; (c) blocking asynchronous send; (d) non-blocking asynchronous.

Source: https://www.css.uic.edu/~ajayk/DCS-Book.

5.6.2. Processor Synchrony

In addition to the distinction between asynchronous and synchronous communication primitives, there is also a distinction between asynchronous and synchronous processors. Processor synchronization denotes that all processors operate in unison, with the synchronization of clocks. Because this synchrony isn't possible in the distributed systems, what is meant more broadly is that these processors are synchronized for significant granularity of code, commonly referred to as a step. This concept is executed by means of some type of boundary synchronization to guarantee that no CPU starts running the next step of code till all processors have finished implementing the previous steps given to every processor (Saglam and Mooney, 2001; Tullsen, Lo, Eggers, and Levy, 1999).

5.6.3. Libraries and Standards

The essential ideas governing all communication primitives were described in the former subsections. Some publicly accessible interfaces that represent most of the above notions are briefly listed in this subsection.

There are numerous primitives for message passing. Several business software products (payroll, banking, and so on) rely on exclusive basic libraries distributed with the vendor's program (for example, the IBM CICS software which has a broadly installed customer base all over the world utilizes its self-primitives). The MPI (message passing interface) library (Blackford et al., 1997; Gropp, 1994) and PVM (parallel virtual machine) library (Sunderam, 1990) are widely used in science, but additional libraries prevail. Commercial software is frequently created utilizing the RPC mechanism (Ananda et al., 1992; Birrell and Nelson, 1984), which allows processes that may reside all across network to be triggered transparently to the user in almost the same way that the local procedure is handled (Ananda et al., 1992; Birrell and Nelson, 1984; Nelson, 1981). To call the method globally, socket primitives or the socket-like transport layer primitives are used behind the scenes. There are numerous RPC executions (Ananda et al., 1992; Campbell et al., 1999; Coulouris et al., 2005), such as Sun RPC and distributed computing environment (DCE) RPC. Other modes of communication include messaging and streaming. Various agencies are proposing and standardizing collections for RMI and ROI (remote object invocation) according to their own pair of primitives as object-based software grows (Campbell et al., 1999). Other established designs according to their own group of primitives include CORBA (Vinoski, 1997) and DCOM (distributed component object model) (Clos, 1953). Various initiatives in research phase are also developing their own flavor of communication primitives.

5.7. SYNCHRONOUS VS. ASYNCHRONOUS EXECUTIONS

Further to the 2 classifications of processor asynchrony/synchrony and asynchronous/synchronous communication primitives, a third category exists, termed as asynchronous/synchronous implementations (He, Josephs, and Hoare, 2015; Motoyoshi, 2004; Rico and Cohen, 2005).

Asynchronous execution occurs when (i) processor synchronization is absent and the rate of drift of the processor clocks is unlimited; (ii) delays in messages (propagation + transmission times) are limited but unbounded; and (iii) there isn't any upper limit on the time it takes for a process to complete a step. Figure 5.13 depicts an instance of asynchronous execution with 4 processes, P_0 through P_3. The messages are represented by arrows, with the transmit and receive events for each message designated by a vertical line and the circle, correspondingly. Shaded circles represent non-communication events, often known as internal events.

A synchronous implementation is the one in which (i) processors are coordinated and the drift rate of clock among any 2 processors is limited; (ii) delivery times of message are that they emerge in one step — or round; and (iii) there is the known upper limit on the time required to implement a step by the process. Figure 5.14 depicts an instance of synchronous execution with 4 processes P_0–P_3. Arrows represent the messages.

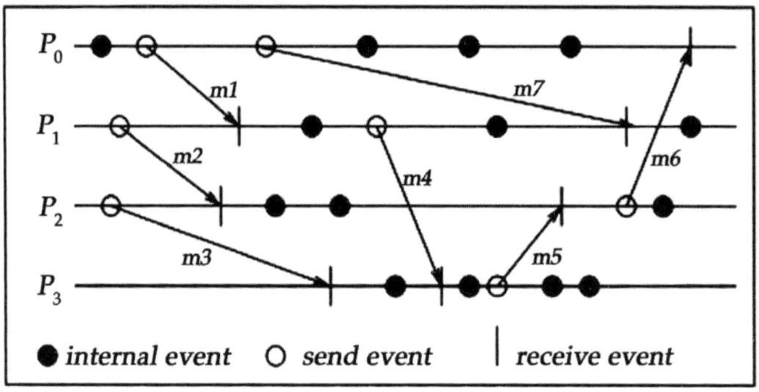

Figure 5.13. In the message-passing system, this is an instance of P_0 asynchronous implementation. The execution is depicted using a P_1 timing diagram, P_2, internal event send and receive event, P_3.

Source: https://dl.acm.org/doi/book/10.5555/1374804.

The synchronized nature of the implementations at all processes makes it simpler to design and confirm algorithms supposing synchronous execution. There is, moreover, a barrier to a genuinely synchronous execution. It is hard to construct a totally synchronous system with messages conveyed within a specified time frame. Consequently, this synchrony must be simulated behind the scenes, which will necessitate delaying or halting certain procedures for a period of time. Therefore, synchronous execution is a concept that must be made available to programs. Notice, when executing this concept, that the very few processor steps or synchronizations, the fairly low the delays and expenses. If processors are permitted to execute asynchronously for a span of time and afterwards, they synchronize, then maybe the synchronization is coarse. In reality, this is nearly synchronous implementation, and the corresponding idea is often referred to as virtual synchrony. Idealistically, numerous programs want procedures to implement a set of instructions in rounds (also known as phases or steps), with the necessity that after every round/phase, all procedures must be coordinated and that all of the sent messages must be delivered. A synchronous execution is generally comprehended in this way. Processes might execute a limited and bounded amount of sequential sub-rounds within each round/phase/step. Every round is supposed to send only 1 message for one process, resulting in the single message hop for the message sent.

Figure 5.14 depicts the timing diagram of the instance of synchronous execution. This system contains 4 nodes P_0 through P_3. Procedure P_i sends the message to P(i+1) mod 4 and P(i1) mod 4 in every iteration and computes an application-specific operation on the received values.

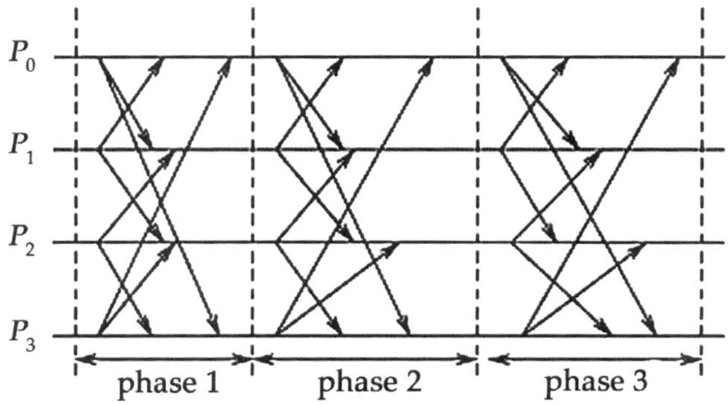

Figure 5.14. An instance of the synchronous implementation in the P_0 message passing system. All of the messages delivered in the round P_1 are received in that particular round.

Source: https://dll.acm.org/doi/book/10.55555/1374804.

5.7.1. Mirroring the Asynchronous System via Synchronous System

Because the synchronous scheme is the specialized case of an asynchronous system, the asynchronous program can be mirrored on the synchronous system fairly easily.

5.7.2. Mirroring the Synchronous System via an Asynchronous System

The synchronous program can usually be mirrored on the asynchronous system utilizing a tool known as *synchronizer*.

5.7.3. Emulations

The message-passing system could emulate a common memory system, and conversely. As depicted in Figure 5.15, there are now 4 broad program categories. Using the displayed emulations, any category can emulate the other. If the system A can typically be mirrored by the system B, denoted as A/B, and an issue can't be solved in B, then it can't be solved in the system A as well. Similarly, if an issue can be solved in A, it can also be solved in B. Consequently, in aspects of computability – what can't and can be computed – all four categories are similar in the systems free of failure.

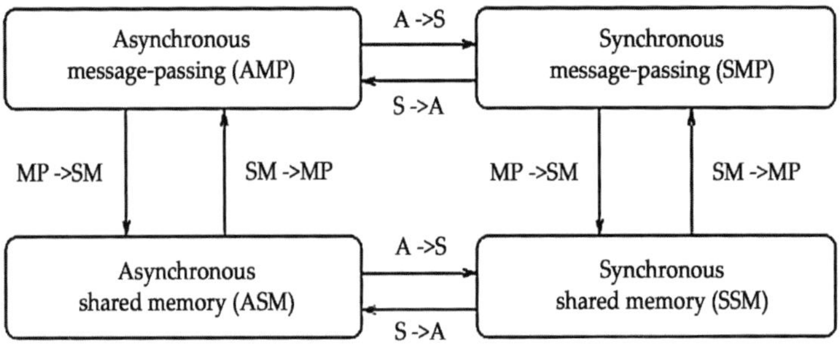

Figure 5.15. Emulations between the principal system categories in the system free of failure.

Source: https://dll.acm.org/doi/book/10.5555/1374804.

5.8. DESIGN PROBLEMS AND CHALLENGES

Since the 1970s, when the ARPANET and internet were developed, distributed computing networks have been widely utilized. Offering access to distant information in the face of malfunctions, file system architecture, and directory structure architecture were the main concerns in the architecture of distributed systems. Though these persist to be critical problems, numerous newer problems have emerged as the rapid expansion of high-speed, high-bandwidth distributed and internet applications has led to the emergence of numerous novel issues (Abbate, 1994; Hauben, 2007; Kirstein, 1999; Perry, Blumenthal, and Hinden, 1988).

The critical design problems and challenges are described below after classifying them as (i) having a greater module associated to the system design and design of operating system; (ii) having the greater module associated to algorithm design; or (iii) evolving from current technological breakthroughs and driven by novel applications. There exists some overlap amongst these classes. Moreover, due to the chasm between the systems community, (a) theoretical algorithms community inside distributed computing, and (b) the forces driving evolving areas and technology, it is beneficial to identify these classes. For instance, existing distributed computing process adheres to the client–server design to a large extent, while this receives little consideration in theoretical distributed algorithms community. The following are reasons for the chasm. First, beyond the scientific computing community, the vast majority of applications that utilize distributed networks are commercial applications for which simplified models are sufficient. For instance, ever since 1970s and 1980s, the customer–server design has been strongly rooted with legacy applications invented by Blue Chip companies (for example, HP, Compaq, IBM, Wang). For most typical commercial applications, this model suffices. Second, industry standards, which don't always select technically best solution, have a large influence on the condition of the practice.

5.8.1. Distributed Systems Challenges From the System Viewpoint

When designing and creating the distributed system, the following functions should be considered (Bittencourt et al., 2018; Iwanicki, 2018; Zhang, Yin, Peng, and Li, 2018).

5.8.1.1. Designing Beneficial Execution Designs and Structures

Two broadly used designs of the distributed system execution are the interleaving and partial order model. They've proven to be especially beneficial for operational explanation and distributed algorithm design. Other designs that offer different extents of set-up for explaining more explicitly with and proving the accurateness of distributed programs include the input/output automata model (Birman, 1993; Lynch, 1996; Peleg, 2000) as well as the TLA (temporal logic of actions).

5.8.1.2. Dynamic Distributed Graph Algorithms and Distributed Routing Algorithms

The distributed system is represented as the distributed graph, with graph algorithms serving as the foundation for a wide range of higher-level communication, data aggregation, object positioning, and object search functionalities. In the routing algorithm, the algorithms must cope with dynamically varying graph features, like modeling altering connection loads. The effectiveness of these algorithms has an effect on not just user-perceived delay, but also traffic and, as a result, network load or overload. As a result, designing effective distributed graph algorithms is critical (Ghosh, 2006).

5.8.1.3. Time and Overall State in the Distributed System

The system's processes are distributed over 3 dimensions of physical space. One more dimension, time, must be appropriately imposed across space. The difficulties stem from supplying precise physical time as well as a time variant known as logical time. This time is comparative time, and it avoids the overhead of supplying physical time for uses that don't need physical time. More crucially, logical time may (i) capture the distributed program's logic and inter-process relationships; as well as (ii) measure the comparative progress at every process.

Monitoring the overall state of a system necessitates the incorporation of the dimension of time for continuous observation. Because of the distributed architecture of the system, it is impossible for a single process to immediately monitor a relevant overall state throughout all processes without coordinating additional state-gathering work (Sunderam, 1990; Wu, 2017).

The time dimension is also involved in determining acceptable concurrency metrics, as evaluating the independence of distinct implementation threads is dependent not just on logic of program but also on execution rates inside logical processes and communication speeds between threads.

5.8.1.4. Synchronization Methods

The processes should be permitted to run synchronously, unless they want to synchronize in order to share information, i.e., interact about common data. Synchronization is required for distributed procedures to circumvent the restricted monitoring of system state from any one process. Avoiding this restricted observation is required before taking any decisions that will have an effect on other procedures. Synchronization methods can also be thought of as resources planning and concurrency management procedures that help to simplify the behavior of procedures that might otherwise function independently. Here are some instances of synchronization problems:

- **Physical Clock Synchronization:** Because of hardware restrictions, the values of physical clocks frequently deviate. Maintaining shared time is a basic problem that requires synchronization.
- **Election of Leader:** All of the processes must agree on which of them will take on the part of a prominent process, also known as the leader process. Even with numerous distributed algorithms, a leader is required since there is typically some imbalance – for example, launching a broadcast or gathering the system's state, or restoring a token that has become lost in the system.
- **Mutual Exclusion:** Since access to the essential resource(s) must be synchronized, this is definitely a synchronization issue.
- **Deadlock Detection and Resolution:** These must be coordinated to prevent duplication of effort and wasteful process terminations, respectively.
- **Termination Detection:** To identify the precise overall state of quiescence, the processes must work together.
- **Garbage Collection:** Garbage is defined as objects that aren't in utilization and aren't referenced by some other process. Identifying garbage necessitates process collaboration.

5.8.1.5. Group Communication, Multicast, and Ordered Message Delivery

A group is an amalgamation of procedures inside the domain of application that part a shared context and cooperate on a mutual task. To allow effective communication and management of group in which processes can enter and leave groups periodically or fail, certain algorithms must be devised. When many processes transmit messages simultaneously, recipients might receive the messages in various orders, thereby defying the distributed program's principles. Consequently, explicit requirements of the principles of ordered delivery must be developed and executed (Birman, Schiper, and Stephenson, 1991; Kihlstrom, Moser, and Melliar-Smith, 1998; Moser et al., 1996).

5.8.1.6. Observing Distributed Events and Bases

Predicates are being used for defining constraints on the overall system state and are important for troubleshooting, monitoring the environment, and industrial process management. They are specified on program parameters that are local to various processes. As a result, online algorithms for assessing such predicates are critical. The event streaming model, in which streams of appropriate events stated by many processes are analyzed together to discover predicates, is a significant paradigm for observing distributed events. Generally, physical, or logical temporal linkages are used to specify such predicates.

5.8.1.7. Distributed Program Design and Confirmation Tools

Programs that are methodically developed and indisputably accurate can significantly minimize the cost of software design, troubleshooting, and development. It is difficult to create mechanisms to fulfill these design and validation aims.

5.8.1.8. Troubleshooting Distributed Programs

Troubleshooting sequential programs is difficult; troubleshooting distributed programs is much more difficult due to parallelism in actions and the ambiguity resulting from a huge array of potential executions specified by the overlapping concurrent operations. To overcome this problem, sufficient debugging procedures and tools must be built.

5.8.1.9. Data Repetition, Uniformity Models, and Caching

Data and other sources must be mirrored in a distributed system for quick access. When looking for duplicates, the issue of assuring coherence among duplicates and stored copies arises. Furthermore, because resources are rarely freely reproduced, placing the duplicates in the system is a challenge.

5.8.1.10. World Wide Web (WWW) Design – Caching, Searching, Scheduling

The web is an instance of a widely dispersed system having a direct interface to the end customer, where most objects' activities are largely read-intensive. The above-mentioned object duplication and caching difficulties must be customized to the web. Prefetching of items is also possible when access trends and other properties of the entities are known. Adhering to Content

Distribution Servers is one instance of how prefetching can be employed. Lowering response time in order to reduce user-perceived delays is a significant task. Item search and browsing on the web are critical functions in the web's operation, yet they consume a lot of resources. It is a significant problem to design methods that can do this effectively and accurately (Haynes and Noveck, 2015; Shirley, 1992).

5.8.1.11. Distributed Common Memory Abstraction

A common memory abstraction facilitates the programer's job since only read and write functions and none of the message communication primitives are required. In middleware layer, meanwhile, the assumption of a common address space must be accomplished by message forwarding. Therefore, the common memory abstraction isn't cheap in terms of expenses (Raynal, 1988; Zaharia et al., 2012).

- **Wait-Free Algorithms:** Wait-freedom, which is considered as a procedure' ability to finish its execution regardless of the activities of other processes, became popular in the creation of algorithms to manage access to common resources in the common memory abstraction. It is an essential characteristic in fault-tolerant systems engineering because it correlates to n 1-mistake adaptability in a n process system. Whereas wait-free algorithms are desired, they are also costly, and inventing wait-free algorithms with low cost is difficult.
- **Shared Exclusion:** A first lesson in operating systems describes the fundamental algorithms for mutual exclusion in the multiprocessing common memory setting (like the Bakery method and the use of semaphores). This book will address more advanced algorithms, like ones centered on hardware primitives, rapid shared exclusion, and wait-free algorithms.
- **Register Constructions:** In light of compelling and innovative technologies of the future – like quantum computing and biocomputing – that have the ability to modify the existing basis of computer "hardware" architecture, we must reconsider the memory admittance suppositions of existing systems that are based purely on semiconductor techniques and the von Neumann design. Particularly, the supposition of single and multiport memory with serial accessibility over the bus in strict

synchronization with the hardware clock of a system might not be applicable when parallel "unlimited" and "overlapping" access to the very same memory region is possible. The analysis of register constructions focuses on the building of registers from start, under the presumption that register accesses are extremely limited. This field serves as the basis for future designs that offer unrestricted access control to even the most elementary memory units (regardless of technology).

- **Consistency Models:** Various degrees of coherence between duplicates of a variable can be permitted when there arc numerous replicas of the variable. These reflect a trade-off between coherence and execution expense. A tight description of consistency would obviously be costly to achieve in terms of delay, message overflow, and concurrency. As a result, models of coherence that are loosened but still significant are preferred.

5.8.2. Load Balancing

The purpose of load balancing is to increase throughput while decreasing user-perceived delay. Load balancing might well be required due to a number of variables, including excessive internet traffic or a high rate of request, which causes the network connection to become a barrier, or higher processing load. Load balancing is frequently employed in datacenters, where the goal is to handle arriving client requests as quickly as possible. Various classic operating system outcomes can be utilized here, but they must be tailored to the requirements of the dispersed context. Load balancing can take the following forms:

- **Data Migration:** The capacity to relocate data (which might be duplicated) within a system centered on user access patterns.
- **Calculation Migration:** The capacity to move processes around in order to redistribute the burden.
- **Distributed Scheduling:** This improves user response time by making greater use of idle computing resources in the system.

5.8.3. Real-Time Scheduling

This is essential for mission-critical systems in order to execute tasks on time. The difficulty of the challenge increases in the distributed system where there isn't an overall view of the system's status. Without such an overall perspective of the status, it is also tougher to make online or periodic modifications to the timetable (Cheriton and Skeen, 1993).

Moreover, network-dependent message transmission latencies are difficult to manage or anticipate, making real-time commitments that are intrinsically reliant on communication across processes more difficult. Though effectiveness of service networks can be deployed, they only help to reduce the unpredictability in latencies to a limited amount. Furthermore, such networks might not be available at all times.

5.8.4. Performance

Though high throughput isn't the main purpose of employing a distributed system, it is vital to have optimal performance. Network delay and access to common resources can cause significant latencies in highly distributed systems, which should be reduced. The user-perceived response time is critical.

The following are few instances of challenges that may emerge when assessing performance:

- **Metrics:** Suitable metrics should be created or found for assessing the performance of both theoretical distributed algorithms and their executions. The earlier would consist of numerous intricacy indicators on metrics, while the later would consist of several system and analytical metrics.
- **Measurement Tools/Methods:** Because a genuine distributed system is a complicated entity that must cope with all of the challenges that come with performance measurement via a WAN or the Internet, adequate methodology and tools for assessing performance metrics should be established.

REFERENCES

1. Abbate, J. E., (1994). *From ARPANET to internet: A history of ARPA-Sponsored Computer Networks, 1966–1988*. University of Pennsylvania.
2. Alboaie, L. *Distributed Systems-Overview*.(pp. 1-25)
3. Ananda, A. L., Tay, B., & Koh, E.-K. (1992). A survey of asynchronous remote procedure calls. *ACM SIGOPS Operating Systems Review, 26*(2), 92–109.
4. Beneš, V. E., (1965). *Mathematical Theory of Connecting Networks and Telephone Traffic*. Academic press.
5. Birman, K. P., (1993). The process group approach to reliable distributed computing. *Communications of the ACM, 36*(12), 37–53.
6. Birman, K., Schiper, A., & Stephenson, P., (1991). Lightweight causal and atomic group multicast. *ACM Transactions on Computer Systems (TOCS), 9*(3), 272–314.
7. Birrell, A. D., & Nelson, B. J., (1984). Implementing remote procedure calls. *ACM Transactions on Computer Systems (TOCS), 2*(1), 39–59.
8. Bittencourt, L. F., Goldman, A., Madeira, E. R., Da Fonseca, N. L., & Sakellariou, R., (2018). Scheduling in distributed systems: A cloud computing perspective. *Computer Science Review, 30*, 31–54.
9. Blackford, L. S., Choi, J., Cleary, A., D'Azevedo, E., Demmel, J., Dhillon, I., & Petitet, A., (1997). *ScaLAPACK Users' Guide*. SIAM.
10. Campbell, A. T., Coulson, G., & Kounavis, M. E., (1999). Managing complexity: Middleware explained. *IT Professional, 1*(5), 22–28.
11. Cheriton, D. R., & Skeen, D., (1993). *Understanding the Limitations of Causally and Totally Ordered Communication.* Paper presented at the proceedings of the fourteenth ACM symposium on operating systems principles.
12. Clos, C., (1953). A study of non-blocking switching networks. *Bell System Technical Journal, 32*(2), 406–424.
13. Cooley, J. W., & Tukey, J. W., (1965). An algorithm for the machine calculation of complex Fourier series. *Mathematics of Computation, 19*(90), 297–301.
14. Coulouris, G. F., Dollimore, J., & Kindberg, T., (2005). *Distributed Systems: Concepts and Design*. Pearson education.

15. Cypher, R., & Leu, E., (1994). *The Semantics of Blocking and Nonblocking Send and Receive Primitives*. Paper presented at the proceedings of 8th international parallel processing symposium.
16. Feng, T. Y., (1981). A survey of interconnection networks. *Computer, 14*(12), 12–27.
17. Flynn, M. J., (1972). Some computer organizations and their effectiveness. *IEEE transactions on Computers, 100*(9), 948–960.
18. Ghosh, S., (2006). *Distributed Systems: An Algorithmic Approach*. Chapman and Hall/CRC.
19. Goscinski, A., (1991). *Distributed Operating Systems: The Logical Design*. Addison-Wesley Longman Publishing Co., Inc.
20. Grama, A., Kumar, V., Gupta, A., & Karypis, G., (2003). *Introduction to Parallel Computing*. Pearson Education.
21. Gropp, W., & Lusk, E., (1994). *Anthony Skjellum: USING MPI Portable Parallel Programming with the Message-Passing Interface*. In: The MIT Press, Cambridge-Massachusetts-London.
22. Gropp, W., (1994). Ewing Lusk, and Anthony Skjellum. *Using MPI: Portable Parallel Programming with the Message Passing Interface*. In: The MIT Press Cambridge, MA.
23. Harary, F., Hayes, J. P., & Wu, H. J., (1988). A survey of the theory of hypercube graphs. *Computers & Mathematics with Applications, 15*(4), 277–289.
24. Hauben, M., (2007). *History of ARPANET* (Vol. 17, pp. 1–20). *Website of l'Instituto Superior de Engenharia do Porto*.
25. Haynes, T., & Noveck, D., (2015). *Network File System (NFS) Version 4 Protocol* (2070-1721)(Vol. 1, pp. 1-5).
26. He, J., Josephs, M. B., & Hoare, C. A. R., (2015). *A Theory of Synchrony and Asynchrony*.(pp.1-4)
27. Iwanicki, K., (2018). *A Distributed Systems Perspective on Industrial IoT*. Paper presented at the 2018 IEEE 38th International Conference on Distributed Computing Systems (ICDCS).
28. Kihlstrom, K. P., Moser, L. E., & Melliar-Smith, P. M., (1998). *The SecureRing Protocols for Securing Group Communication*. Paper presented at the proceedings of the thirty-first Hawaii international conference on system sciences.

29. Kirstein, P. T., (1999). Early experiences with the Arpanet and Internet in the United Kingdom. *IEEE Annals of the History of Computing, 21*(1), 38–44.
30. Lea, D., Vinoski, S., & Vogels, W., (2006). Guest editors' introduction: Asynchronous middleware and services. *IEEE Internet Computing, 10*(1), 14–17.
31. Lynch, N. A., (1996). *Distributed Algorithms*. Elsevier.
32. Moser, L. E., Melliar-Smith, P. M., Agarwal, D. A., Budhia, R. K., & Lingley-Papadopoulos, C. A., (1996). Totem: A fault-tolerant multicast group communication system. *Communications of the ACM, 39*(4), 54–63.
33. Motoyoshi, I., (2004). The role of spatial interactions in perceptual synchrony. *Journal of Vision, 4*(5), 1–1.
34. Nelson, B. J., (1981). *Remote Procedure Call.* Carnegie Mellon University.
35. Peleg, D., (2000). *Distributed Computing: A Locality-Sensitive Approach*. SIAM.
36. Perry, D. G., Blumenthal, S. H., & Hinden, R. M., (1988). *The ARPANET and the DARPA Internet.* Library Hi Tech.
37. Raynal, M., (1988). *Distributed Algorithms and Protocols*. Chichester: Wiley.
38. Rico, R., & Cohen, S. G., (2005). Effects of task interdependence and type of communication on performance in virtual teams. *Journal of Managerial Psychology*.
39. Saglam, B. E., & Mooney, V. J., (2001). *System-on-a-Chip Processor Synchronization Support in Hardware.* Paper presented at the proceedings design, automation, and test in Europe. Conference and exhibition 2001.
40. Shirley, J., (1992). *Guide to Writing DCE Applications*: O'Reilly & Associates, Inc.
41. Singhal, M., & Shivaratri, N. G., (1994). *Advanced Concepts in Operating Systems*. McGraw-Hill, Inc.
42. Snir, M., Gropp, W., Otto, S., Huss-Lederman, S., Dongarra, J., & Walker, D., (1998). *MPI--the Complete Reference: The MPI Core* (Vol. 1). MIT press.

43. Sunderam, V. S., (1990). PVM: A framework for parallel distributed computing. *Concurrency: Practice and Experience, 2*(4), 315–339.
44. Tullsen, D. M., Lo, J. L., Eggers, S. J., & Levy, H. M., (1999). *Supporting fine-Grained Synchronization on a Simultaneous Multithreading Processor.* Paper presented at the proceedings Fifth international symposium on high-performance computer architecture.
45. Van, S. M., & Tanenbaum, A., (2002). Distributed systems principles and paradigms. *Network, 2*, 28.
46. Vinoski, S., (1997). CORBA: Integrating diverse applications within distributed heterogeneous environments. *IEEE Communications Magazine, 35*(2), 46–55.
47. Wu, C. L., & Feng, T. Y., (1980). On a class of multistage interconnection networks. *IEEE Transactions on Computers, 100*(8), 694–702.
48. Wu, J., (2017). *Distributed System Design*. CRC press.
49. Zaharia, M., Chowdhury, M., Das, T., Dave, A., Ma, J., McCauly, M., & Stoica, I., (2012). *Resilient Distributed Datasets: A {Fault-Tolerant} Abstraction for {in-Memory} Cluster Computing.* Paper presented at the 9th USENIX Symposium on Networked Systems Design and Implementation (NSDI 12).
50. Zhang, Z., Yin, L., Peng, Y., & Li, D., (2018). *A Quick Survey on Large Scale Distributed Deep Learning Systems*. Paper presented at the 2018 IEEE 24th international conference on parallel and distributed systems (ICPADS).

CHAPTER 6

APPLICATIONS OF CONCURRENT, PARALLEL, AND DISTRIBUTED COMPUTING

CONTENTS

6.1. INTRODUCTION

Simultaneous programs are programs that leverage application-level synchronization. Computer operating systems offer three fundamental ways of developing concurrent applications (Sunderam, 1990).

Parallel computing technologies are typically widespread. A parallel system is made up of several units that interact with one another via memory space. Multiprocessor chips would become more popular as the number of transistors on a single chip rise. With adequate parallelism in tasks, these systems would significantly outperform serial systems in terms of efficiency. Such CPUs interact with one another through shared memory. Every CPU might have its private memory which is not accessed by some other CPUs (Kumar, 2005). We describe distributed applications as computer networks that have numerous CPUs linked together through a communications system. Processors in such systems interact with one another through network communications. Due to the obvious lower costs of computer processors as well as the high-bandwidth networks that link them, these technologies are becoming more widely accessible. The communications system shown in the diagram might be a local network like Ethernet or a public network like the web (Waldspurger et al., 1992).

6.2. APPLICATIONS OF CONCURRENT COMPUTING

6.2.1. An Elevator System

We shall have used elevator system terms to demonstrate the ideas. More specifically, we refer to a computer network intended to operate several lifts in a single place inside a structure. Numerous things could be happening at the same time inside a group of lifts almost nothing! Somebody on any level might demand a lift at any moment, and other demands may well be waiting. Many elevators may indeed be idle, whereas others are transporting people, responding to a call, or even both (Sunderam et al., 1994). Doors should be opened and closed at the proper times. Individuals could be blocking doors, hitting door open or shut buttons, or picking levels before making up their minds. Screens must be updated, motors must be operated, and so on, all while being monitored by the lift control scheme. Ultimately, it's a suitable model for investigating parallelism concerns, and another for that we have a reasonable level of knowledge and a workable language (Figure 6.1) (Goudarzi et al., 2020).

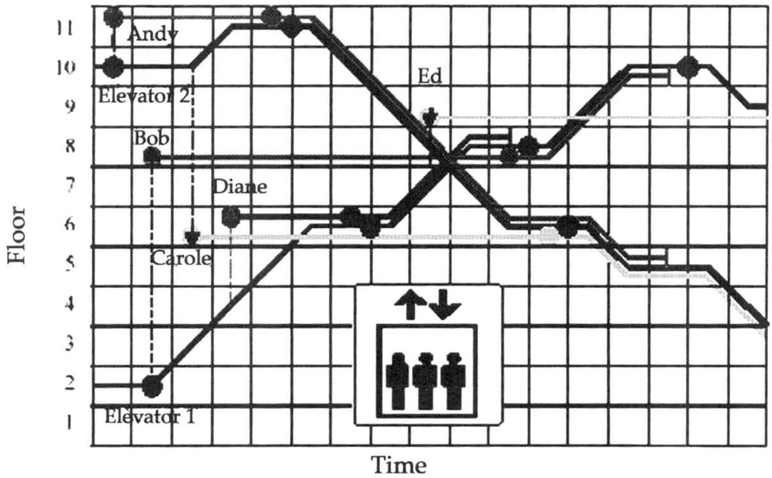

Figure 6.1. A scenario involving two elevators and five potential passengers distributed over 11 floors.

Source: https://swi.cs.vsb.cz/RUPLarge/core.base_rup/guidances/concepts/concurrency_EE2E011A.html.

6.2.2. Multitasking Operating System

The activity is a general factor of parallelism whenever the operating system supports multiplexing. A procedure is an object that the operative system creates, supports, and manages with the express aim of providing an atmosphere upon which to run a program. The procedure gives a memory area for the app program's sole usage, an activity thread for running it, and maybe a few mechanisms for sending and receiving data from other tasks. The activity acts as a virtual CPU for running several threads of an application at the same time (Jia et al., 2014).

Every process may exist in one of three phases (Jean et al., 1999):

- **Blocked:** Waiting for a response or to gain the power of a resource.
- **Ready:** Awaiting a round to operate from the computer system.
- **Running:** The CPU is being used.

Comparative priority is often ascribed to procedures. The kernel of the operating system chooses which processes to execute anywhere at a given moment depending on their conditions, objectives, as well as some kind of scheduled strategy. All the activities in multitasking computer systems are controlled by a single thread (Buyya et al., 2013).

6.2.3. Multithreading

Specific operating systems, especially those designed for practical systems, provide "threads" or "lightweight threads" as a "reduced weight" option for tasks. Threads allow you to get a greater degree of parallelism inside a program. Every thread is associated with a single activity and all strands within this procedure use the same memory space (De Luca and Chen, 2019).

6.2.4. Multiprocessing

Having many processors allows for true concurrent execution. Many tasks are allocated to a certain proccssor forever. However, in other cases, work might be temporarily allocated to the next accessible CPU. Using a "symmetric multiprocessor" is probably the most accessible method to achieve this. Multiple CPUs may access memory over a shared bus in this hardware setup. Threads may be dynamically assigned to any accessible CPU in computer systems that enable symmetric multiprocessors. SUN's Solaris and Microsoft's Windows NT are two cases of active systems that enable symmetric multiprocessors (Geist and Sunderam, 1992).

6.2.5. Function of Wrist-Watch

The installation of any DSM setting roughly mirrors the creation of the watch instance. The issuing domain, the target platform, and the DSM environment are always present in this kind of scenario. Because the domain and target platform are already in place, the goal of creating a DSM environment is to identify their important elements and build mappings between them (Geist and Sunderam, 1992).

This method of dismantling the watch results in a high level of reusability. Logical applications and displays may be developed independently from one another as sub-applications, and they interact through pre-defined interfaces, making them natural components. This allows the developer to swiftly create new watch versions by integrating sub-apps into new logical watch applications, which can then be combined with displays (new or existing). Figure 6.2 depicts this concept of watch setup (Virding et al., 1996).

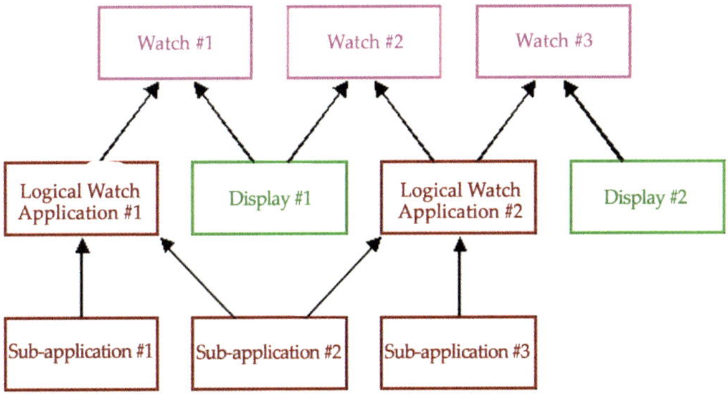

Figure 6.2. How to set up your watch.

Source: https://www.metacase.com/support/40/manuals/watch40sr2a4.pdf.

6.3. APPLICATIONS OF PARALLEL COMPUTING

6.3.1. Science, Research, and Energy

Thanks to parallel processing, you can watch the weather predictions on your smartphone using the Weather Channel app. Not so much because your phone is operating many apps at the same time — parallel computing isn't the same as concurrent computing — but due to maps of climate patterns needing parallel processing (Wende et al., 2014).

Other current studies, such as astrophysics models, seismic surveys, quantum chromodynamics, and more, rely on parallel computing. Let's look more closely at some of them (Navarro et al., 2014).

6.3.1.1. Black Holes and Parallel Supercomputers

Astronomy is a slow-moving science. Stars colliding, galaxies merging, and black holes swallowing celestial objects may consume millions of years, and that is why cosmologists have to use computer models to investigate these phenomena. And the like complicated models need a lot of computing power (Bjørstad and Mandel, 1991).

A parallel supercomputer, for instance, was responsible for the latest breakthrough in the research of black holes. Researchers have shown that the deepest component of stuff that circles and ultimately falls into black holes coincides with all those black holes, solving a four-decade-old enigma. This is critical in aiding scientists' understanding of how this still-mysterious phenomenon operates (Figure 6.3) (Rajasekaran and Reif, 2007).

Figure 6.3. Parallel supercomputers.

Source: https://phys.org/news/2018-06-supercomputer-astronomy.html.

"These characteristics surrounding the black hole might seem insignificant, yet they have a huge influence about what occurs in the galaxy overall," said Northwestern University researcher Alexander Tchekhovskoy, who collaborated on the work alongside the Universities of Amsterdam and Oxford. "They possess power over how rapidly black holes spin as well as, as a consequence, how black holes affect their whole galaxies" (Padua, 2011).

6.3.1.2. Oil and Gas Industry

Bubba, among the top players in the oil sector, stays in a suburb of Houston. But Bubba isn't a black-gold tycoon; it's an Australian geospatial firm DownUnder GeoSolutions' supercomputer (one of the fastest on the globe). Seismic data analysis has long been used to assist create a better image of subterranean strata, which is essential for businesses such as oil and natural gas (Culler et al., 1997). However, numerical computation is now essentially a standard practice in energy exploration, particularly when algorithms analyze large volumes of data to aid drillers in tough terrain such as salt domes. (Total, the largest and most powerful industrial supercomputer, is used by the French energy giant.) Millions of Intel Xeon Phi multiprocessors are used in Bubba's backbone, which is chilled in chilled oiled baths, allowing for extremely high-performance parallelization. By offering parallel electricity related to various enterprises, the expectation is that fewer energy providers would feel driven to create their own, inefficient networks (Rashid et al., 2018).

6.3.1.3. Agriculture and Climate Patterns

Each month, the United States Department of Agriculture assesses the supply side for a variety of main crops. Predictions are critical for everybody, from politicians trying to fix marketplaces to farmers looking to handle their money. The year before, scientists at the University of Illinois' Department of Natural Resources and Environmental Sciences outperformed the federal government's manufacturing estimation by integrating more information—crop development analyzes, seasonal climatic variables, and satellite measurements data — which was then squeezed utilizing machine learning (ML) algorithms on the university's concurrent computing, the petascale Blue Waters. Their projection was shortly shown to be roughly five bushels per hectare more correct. In 2019, the group turned its predicting focus to Australian wheat yields, and the findings were as remarkable (Figure 6.4) (Ferrari, 1998).

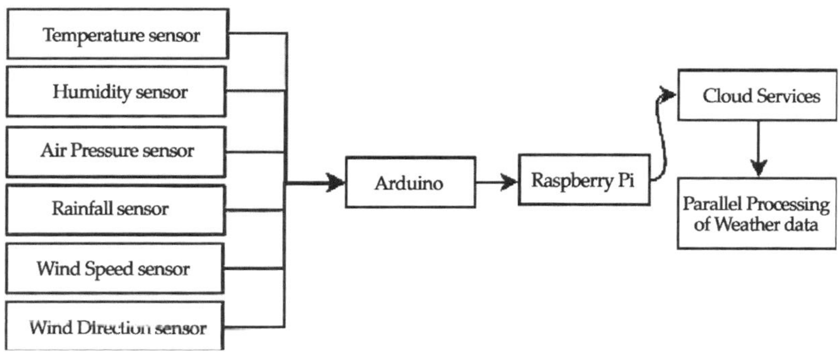

Figure 6.4. Development of a weather monitoring program.

Source: https://www.semanticscholar.org/paper/Parallel-Processing-Implementation-on-Weather-for-Susanto-Sukoco/e2a9e6df57c6b38a1dd98b0a12bea4da90db45cc/figure/0.

6.3.2. The Commercial World

Even though parallel computing is frequently the realm of various official research universities, the business world has taken attention. "The banking sector, investing industry traders, bitcoin — they are the major groups that use a bunch of GPUs to make money," they explained (Ekanayake and Fox, 2009).

Parallel computing has its origins in the show business, which is not surprising considering that GPUs were originally developed for large graphical demands. It is also beneficial to companies that depend on fluid mechanics, and physical research with numcrous commercial uses. Take a closer look (Konfrst, 2004).

6.3.2.1. Banking Sector

Almost every significant facet of modern banking is GPU-accelerated, from credit ratings to hazard modeling to identity theft. In some ways, the shift away from conventional CPU-powered research was unavoidable. GPU offloading matured in 2008, just as politicians passed a slew of post-crash economic laws. In 2017, Xcelerit co-founder Hicham Lahlou stated to The Next Platform, "It's not unusual to see a bank with thousands and thousands of Teslas GPUs." "And it would not have been the situation with no regulatory drive" (Satish et al., 2012).

JPMorgan Chase was an early user, announcing in 2011 that switching from CPU-only to GPU-CPU hybrid computing increased risk calculations by 40% as well as saved the company money by 80%. Wells Fargo recently installed Nvidia GPUs for procedures ranging from speeding AI models for liquidity management to remotely accessing its pc architecture. GPUs were also at the core of the very latest financial fashion: the crypto-mining frenzy. However, chip prices have subsequently steadied after that specific boom and fall (Gourdain et al., 2009).

6.3.3. Medicine and Drug Discovery

Emerging technology is transforming the medical environment in a range of methods, from virtual reality to treat macular degeneration to advances in biomaterials organs and tissue to the several ways Amazon is set to affect healthcare. Parallel computing has always been present in this field, but it is now primed to generate even more advancements. This is how (Garland et al., 2008).

6.3.3.1. Medical Imaging

Medical imaging was among the first sectors to see a paradigm shift as a result of parallel processing, specifically the GPU-for-general-computing explosion. There is now an entire amount of medical journals detailing how high computing and capacity contributed to enormous increases in velocity as well as clarity for, well, almost everything: MRI, CT, X-rays, optical tomography, etc. The very next breakthrough in medical imaging would almost certainly be parallel-focused, with Nvidia just at the vanguard (Marowka, 2007). Radiologists may more readily obtain the potential of AI by utilizing the company's newly launched toolset, which enables imaging devices to manage to expand volumes of data and computing weight. Clara, the GPU-powered technology, apparently allows clinicians to construct imaging models with 10 times less information than would normally be necessary. Ohio State University as well as the National Institutes of Health are one of the universities which have already signed on (Asanovic et al., 2008).

6.3.3.2. Simulations of Molecular Dynamics

Consider parallel processing to be a nested doll, with one of the deepest figurines being an existing medicine. Parallel programming is a suitable architecture for executing molecular dynamics, which have shown to be extremely beneficial in drug development. Accellera, a clinical research group, has created many applications that make use of GPUs' sophisticated offloading architecture, including the modeling code ACEMD as well as the Python module HTMD. They've been enough to conduct models on some very capable supercomputers, such as a Titan run that enabled researchers better grasp how our neurotransmitter interacts. Accellera has also collaborated on pharmaceutical research with companies such as Janssen and Pfizer. Because sophisticated parallel computing enables fine-grained analysis of molecular machinery, it might have significant applications in the study of hereditary illness, which researchers are actively investigating (Takefuji, 1992).

6.3.3.3. Public Health

Parallel processing's tremendous data-analytics capacity has considerable potential for public health across picture restoration and pharmacological study. Consider one especially heinous epidemic: veteran suicide. As per statistics from the Department of Veterans Affairs, approximately 20 veterans have recently passed away per day since 2014. The subject is just one of those that have received actual bipartisan concern (Sorel, 1994).

Scientists at the Oak Ridge National Laboratory were capable of executing the program on a greater computer 3-fold quicker than the VA's abilities after the VA constructed a framework that delved into veterans' preferences and renewal habits. The goal is to someday deploy IBM's famous (and GPU-packed) Summit computer to provide physicians with real-time danger warnings (Martino et al., 1994).

6.4. APPLICATIONS OF DISTRIBUTED COMPUTING

6.4.1. Mobile Systems

Wireless communication, which would be dependent on electromagnetic waves as well as uses a common broadcast channel, is often used in mobile applications. As a result, communication properties change; several concerns, like radio range as well as capacity, as well as many technical challenges, like battery power saving, interface with wired Web, data processing, as well as interference, come into the equation. From a computer engineering standpoint, there are several issues to consider, including routing, location management, channel access, translation, and location prediction, as well as general mobility management (Figure 6.5) (Siddique et al., 2016).

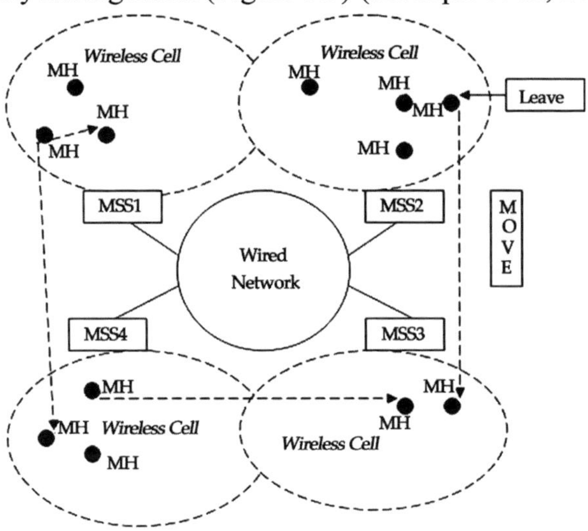

Figure 6.5. The schematic diagram for distributed mobile systems.

Source: https://www.researchgate.net/figure/System-Model-for-Distributed-Mobile-Systems_fig1_271608403.

A mobile network may be built in one of two ways. The first is the base-station strategy, sometimes recognized as the cellular method, where a layer is defined as a geographical area within the scope of a static yet strong base network providing. The base station connects all mobile activities in that unit to the entire system (Fox, 1989; Hoefler and Lumsdaine, 2008). The ad-hoc fully-connected-layer, where there's no network device, is the second important choice acted as a centralized node for its cell). The mobile nodes share complete blame for connectivity, and they must engage in networking by forwarding the packets from other sets of network entities. This is, without a doubt, a complicated model. From a computer science standpoint, it presents several graph-theoretical issues, as well as a variety of engineering challenges (Asanovic et al., 2009).

6.4.2. Sensor Networks

A sensor is a computer that can sense physical factors like temperature, velocity, pressure, humidity, as well as chemicals via an electro-mechanical interface. New progress in low-cost hardware technologies has enabled the deployment of very massive (on the scale of 106 or more) low-cost detectors (Catanzaro et al., 2010). The event streaming paradigm, which has been previously developed, is an essential paradigm for tracking dispersed events. The occurrences recorded by sensor nodes vary from the reported incidents by "computer processes" although the occurrences recorded by a sensor node happen in the real world, outside of the computer system as well as procedures. In a sensor network, this restricts the kind of details relating to the recorded occurrence (Asanovic et al., 2006).

6.4.3. Ubiquitous or Pervasive Computing

Ubiquitous systems are a kind of technology in which computers implanted in and permeating the environment conduct secondary app operations, much like in science fiction movies. The smart home, as well as smart office, are two examples of ubiquitous settings that are now undergoing extensive research for development (Rashid et al., 2019). Ubiquitous computers are fundamentally dispersed systems that make use of wireless technology, as well as sensing and actuation systems. They may be energy-restricted yet self-organizing as well as network-centric. These systems are often described as reaching a large number of tiny computers working together in a fluid contextual network. To manage and collate data, the processors may well be linked to more capable networks as well as processor speed in the backend (Wood and Hill, 1995).

6.4.4. Peer-to-Peer Computing

Peer-to-peer (P2P) computing refers to processing via application-level networking in which all transactions between computers take place on a "peer" level, with no hierarchy between them. As a result, all processors are treated the same and perform a symmetrical part in the calculation. Client-server computing, in which the responsibilities amongst some of the CPUs are largely unequal, gave way to peer-to-peer computing (Barney, 2010; Huck and Malony, 2005). P2P networks are often self-organizing, with or without a regular network topology. For nameserver as well as item lookup, no centralized databases (which are often used in domain controllers) are permitted. Object storage methods, efficient object search online, and retrieval in a systemic way; dynamic recompilation to endpoints as well as items arbitrarily going to join as well as abandoning the network; recombination strategy is a strategy to speed up object lookup; tradeoffs among both object volume delay as well as table sizes; anonymity, confidentiality, and safety are among the critical problems (Figure 6.6) (Hendrickson and Kolda, 2000).

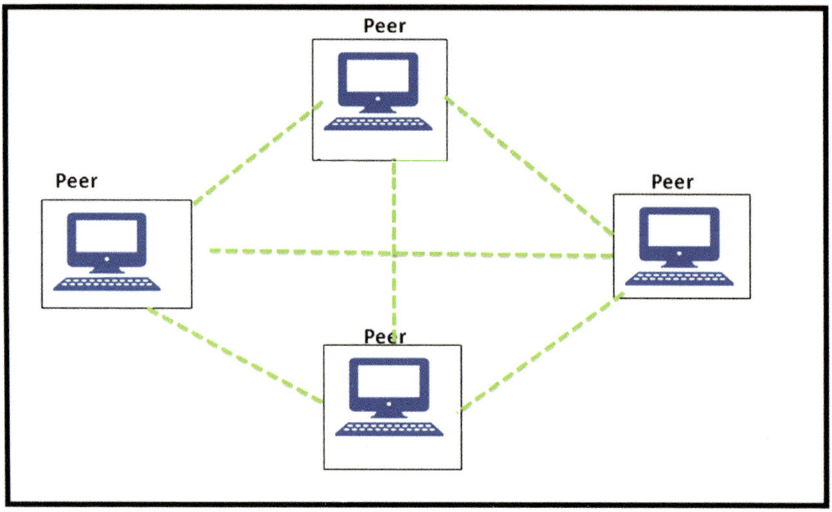

Figure 6.6. Peer-to-peer network.

Source: https://t4tutorials.com/peer-to-peer-network-characteristics-advantages-disadvantages/.

6.4.5. Publish-Subscribe and Multimedia

With the increase in data, there is a growing necessity to acquire as well as analyze only key data. Filters may be used to specify this data. In a changing situation in which things are constantly changing (changing stock prices, for instance), there must be I an effective method for delivering such a data (publish), (ii) an effective method for allowing end-users to imply interest in obtaining relevant data (subscribe), and (iii) an effective method for collating high quantities of information published as well as sorting it according to the user's membership filter (Keckler et al., 2011).

6.4.6. Distributed Agents

Agents are software programs or robotics that may move about the network and do certain activities that they have been designed to accomplish. The term "agent" comes from the notion that agents labor on behalf of a larger goal. Agents acquire and analyze data and may share it with the other agents. The agents frequently collaborate, as in an ant colony, but they could also compete in a free market system. Coordination methods amongst agents, managing agent mobility, plus design phase, as well as interfacing, are all difficulties in distributed applications. Agent study spans machine intelligence, mobile computing, economic market models, software engineering, as well as distributed computing (Li and Martinez, 2005).

6.4.7. Distributed Data Mining

Data mining algorithms mine or derive meaning from massive volumes of data by detecting the changes as well as patterns in the data. Analyzing client purchase habits to profile customers and improve the effectiveness of directed marketing strategies is a typical case. A data reservoir may be mined using databases as well as artificial intelligence (AI) algorithms. Several times, data must be disseminated and could be gathered in a repository, such as in banking apps where other data is confidential and confidential, or in climatic weather forecasting in which the sets of data are simply too large to gather and evaluate in real-time at a single repository. Efficient distributed data mining methods are necessary for such instances (Vrugt et al., 2006).

6.4.8. Grid Computing

The data as well as computing grid, similar to the electrical transmission and distribution grid, is expected to become a fact soon. Briefly said, idle CPU units of network-connected devices would be accessible to others. Numerous difficulties in trying to make grid computing an actuality involve planning jobs in a distributed manner, developing a framework to implement service quality as well as real-time assurances, and, of course, maintaining the safety of physical devices and the jobs being performed in this environment (Figure 6.7) (Lee and Hamdi, 1995).

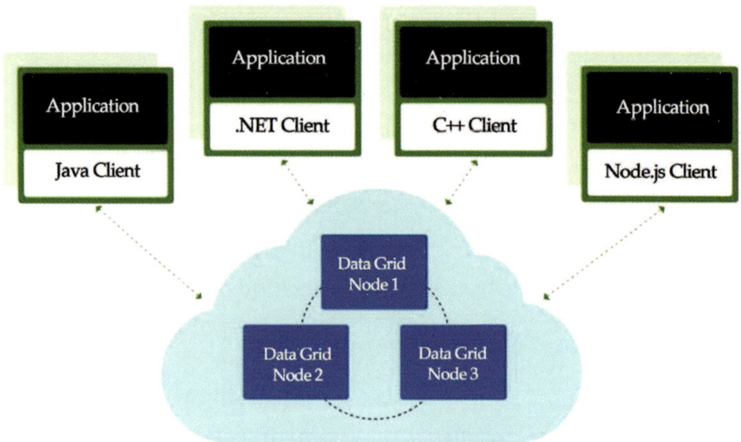

Figure 6.7. Grid computing.

Source: https://hazelcast.com/glossary/grid-computing/.

6.4.9. Security in Distributed Systems

Traditional distributed security problems involve confidentiality (making sure that only approved operations may see particular information), authentication (making sure the origin of information received as well as the integrity of the transmitting operation), and reliability (maintaining a permitted approach to services despite malicious actions) (Gunarathne et al., 2013). The objective is to find effective and useful solutions to such problems. Traditional dispersed setups have solved these fundamental difficulties. Such challenges become much more exciting for the newer architectures mentioned in this sub-section, like the wireless, peer-to-peer, grid, and pervasive computing, owing to circumstances including resource-constrained surroundings resource-constrained surroundings, and a broadcast medium, an absence of structure, and a loss of trust in the system (McColl, 1993).

REFERENCES

1. Asanovic, K., Bodik, R., Catanzaro, B. C., Gebis, J. J., Husbands, P., Keutzer, K., & Yelick, K. A., (2006). *The Landscape of Parallel Computing Research: A View from Berkeley* (Vol. 1, pp. 1–4).
2. Asanovic, K., Bodik, R., Demmel, J., Keaveny, T., Keutzer, K., Kubiatowicz, J. D., & Yelick, K. A., (2008). *The Parallel Computing Laboratory at UC Berkeley: A Research Agenda Based on the Berkeley View* (Vol. 1, pp. 1–6). EECS Department, University of California, Berkeley, Tech. Rep.
3. Asanovic, K., Bodik, R., Demmel, J., Keaveny, T., Keutzer, K., Kubiatowicz, J., & Yelick, K., (2009). A view of the parallel computing landscape. *Communications of the ACM, 1, 52*(10), 1–9, 56–67.
4. Barney, B., (2010). *Introduction to Parallel Computing* (Vol. 6, No. 13, p. 10). Lawrence Livermore National Laboratory.
5. Bjørstad, P. E., & Mandel, J., (1991). On the spectra of sums of orthogonal projections with applications to parallel computing. *BIT Numerical Mathematics, 31*(1), 76–88.
6. Buyya, R., Vecchiola, C., & Selvi, S. T., (2013). *Mastering Cloud Computing: Foundations and Applications Programming* (Vol. 1, pp. 1–3). Newnes.
7. Catanzaro, B., Fox, A., Keutzer, K., Patterson, D., Su, B. Y., Snir, M., & Chafi, H., (2010). Ubiquitous parallel computing from Berkeley, Illinois, and Stanford. *IEEE Micro, 30*(2), 41–55.
8. Culler, D. E., Arpaci-Dusseau, A., Arpaci-Dusseau, R., Chun, B., Lumetta, S., Mainwaring, A., & Wong, F., (1997). Parallel computing on the Berkeley now. In: *9th Joint Symposium on Parallel Processing* (Vol. 1, pp. 1, 2).
9. De Luca, G., & Chen, Y., (2019). Semantic analysis of concurrent computing in decentralized IoT and robotics applications. In: *2019 IEEE 14th International Symposium on Autonomous Decentralized System (ISADS)* (Vol. 1, pp. 1–8). IEEE.
10. Ekanayake, J., & Fox, G., (2009). High performance parallel computing with clouds and cloud technologies. In: *International Conference on Cloud Computing* (Vol. 1, pp. 20–38). Springer, Berlin, Heidelberg.
11. Ferrari, A., (1998). JPVM: Network parallel computing in Java. *Concurrency: Practice and Experience, 10*(11–13), 985–992.

12. Fox, G. C., (1989). Parallel computing comes of age: Supercomputer level parallel computations at Caltech. *Concurrency: Practice and Experience, 1*(1), 63–103.
13. Garland, M., Le Grand, S., Nickolls, J., Anderson, J., Hardwick, J., Morton, S., & Volkov, V., (2008). Parallel computing experiences with CUDA. *IEEE Micro, 28*(4), 13–27.
14. Geist, G. A., & Sunderam, V. S., (1992). Network-based concurrent computing on the PVM system. *Concurrency: Practice and Experience, 4*(4), 293–311.
15. Goudarzi, M., Wu, H., Palaniswami, M., & Buyya, R., (2020). An application placement technique for concurrent IoT applications in edge and fog computing environments. *IEEE Transactions on Mobile Computing, 20*(4), 1298–1311.
16. Gourdain, N., Gicquel, L., Staffelbach, G., Vermorel, O., Duchaine, F., Boussuge, J. F., & Poinsot, T., (2009). High performance parallel computing of flows in complex geometries: II. applications. *Computational Science & Discovery, 2*(1), 015004.
17. Gunarathne, T., Zhang, B., Wu, T. L., & Qiu, J., (2013). Scalable parallel computing on clouds using Twister4Azure iterative MapReduce. *Future Generation Computer Systems, 29*(4), 1035–1048.
18. Hendrickson, B., & Kolda, T. G., (2000). Graph partitioning models for parallel computing. *Parallel Computing, 26*(12), 1519–1534.
19. Hoefler, T., & Lumsdaine, A., (2008). Message progression in parallel computing-to thread or not to thread?. In: *2008 IEEE International Conference on Cluster Computing* (Vol. 1, pp. 213–222). IEEE.
20. Huck, K. A., & Malony, A. D., (2005). Perfexplorer: A performance data mining framework for large-scale parallel computing. In: *SC'05: Proceedings of the 2005 ACM/IEEE Conference on Supercomputing* (Vol. 1, pp. 41–41). IEEE.
21. Jean, J. S., Tomko, K., Yavagal, V., Shah, J., & Cook, R., (1999). Dynamic reconfiguration to support concurrent applications. *IEEE Transactions on Computers, 48*(6), 591–602.
22. Jia, M., Cao, J., & Yang, L., (2014). Heuristic offloading of concurrent tasks for computation-intensive applications in mobile cloud computing. In: *2014 IEEE Conference on Computer Communications Workshops (INFOCOM WKSHPS)* (Vol. 1, pp. 352–357). IEEE.

23. Keckler, S. W., Dally, W. J., Khailany, B., Garland, M., & Glasco, D., (2011). GPUs and the future of parallel computing. *IEEE Micro, 31*(5), 7–17.
24. Konfrst, Z., (2004). Parallel genetic algorithms: Advances, computing trends, applications and perspectives. In: *18th International Parallel and Distributed Processing Symposium, 2004 Proceedings* (Vol. 1, p. 162). IEEE.
25. Kumar, V., (2005). Parallel and distributed computing for cybersecurity. *IEEE Distributed Systems Online, 6*(10), 2–6.
26. Lee, C. K., & Hamdi, M., (1995). Parallel image processing applications on a network of workstations. *Parallel Computing, 21*(1), 137–160.
27. Li, J., & Martinez, J. F., (2005). Power-performance considerations of parallel computing on chip multiprocessors. *ACM Transactions on Architecture and Code Optimization (TACO), 2*(4), 397–422.
28. Marowka, A., (2007). Parallel computing on any desktop. *Communications of the ACM, 50*(9), 74–78.
29. Martino, R. L., Johnson, C. A., Suh, E. B., Trus, B. L., & Yap, T. K., (1994). Parallel computing in biomedical research. *Science, 265*(5174), 902–908.
30. McColl, W. F., (1993). General purpose parallel computing. *Lectures on Parallel Computation, 4*(1), 337–391.
31. Navarro, C. A., Hitschfeld-Kahler, N., & Mateu, L., (2014). A survey on parallel computing and its applications in data-parallel problems using GPU architectures. *Communications in Computational Physics, 15*(2), 285–329.
32. Padua, D., (2011). *Encyclopedia of Parallel Computing* (Vol. 1, pp. 1–7). Springer Science & Business Media.
33. Rajasekaran, S., & Reif, J., (2007). *Handbook of Parallel Computing: Models, Algorithms and Applications* (Vol. 1, pp. 2–4). CRC press.
34. Rashid, Z. N., Zebari, S. R., Sharif, K. H., & Jacksi, K., (2018). Distributed cloud computing and distributed parallel computing: A review. In: *2018 International Conference on Advanced Science and Engineering (ICOASE)* (Vol. 1, pp. 167–172). IEEE.
35. Rashid, Z. N., Zeebaree, S. R., & Shengul, A., (2019). Design and analysis of proposed remote controlling distributed parallel computing system over the cloud. In: *2019 International Conference on Advanced Science and Engineering (ICOASE)* (Vol. 1, pp. 118–123). IEEE.

36. Satish, N., Kim, C., Chhugani, J., Saito, H., Krishnaiyer, R., Smelyanskiy, M., & Dubey, P., (2012). Can traditional programming bridge the ninja performance gap for parallel computing applications?. In: *2012 39th Annual International Symposium on Computer Architecture (ISCA)* (Vol. 1, pp. 440–451). IEEE.
37. Siddique, K., Akhtar, Z., Yoon, E. J., Jeong, Y. S., Dasgupta, D., & Kim, Y., (2016). Apache Hama: An emerging bulk synchronous parallel computing framework for big data applications. *IEEE Access, 4*(2), 8879–8887.
38. Sorel, Y., (1994). Massively parallel computing systems with real time constraints: The "algorithm architecture adequation" methodology. In: *Proceedings of the First International Conference on Massively Parallel Computing Systems (MPCS) The Challenges of General-Purpose and Special-Purpose Computing* (Vol. 1, pp. 44–53). IEEE.
39. Sunderam, V. S., (1990). PVM: A framework for parallel distributed computing. *Concurrency: Practice and Experience, 2*(4), 315–339.
40. Sunderam, V. S., Geist, G. A., Dongarra, J., & Manchek, R., (1994). The PVM concurrent computing system: Evolution, experiences, and trends. *Parallel Computing, 20*(4), 531–545.
41. Takefuji, Y., (1992). *Neural Network Parallel Computing* (Vol. 164, pp. 2–6). Springer Science & Business Media.
42. Virding, R., Wikström, C., & Williams, M., (1996). *Concurrent Programming in ERLANG* (Vol. 1, pp. 1–8). Prentice Hall International (UK) Ltd.
43. Vrugt, J. A., Nualláin, B. O., Robinson, B. A., Bouten, W., Dekker, S. C., & Sloot, P. M., (2006). Application of parallel computing to stochastic parameter estimation in environmental models. *Computers & Geosciences, 32*(8), 1139–1155.
44. Waldspurger, C. A., Hogg, T., Huberman, B. A., Kephart, J. O., & Stornetta, W. S., (1992). Spawn: A distributed computational economy. *IEEE Transactions on Software Engineering, 18*(2), 103–117.
45. Wende, F., Cordes, F., & Steinke, T., (2012). On improving the performance of multi-threaded CUDA applications with concurrent kernel execution by kernel reordering. In: *2012 Symposium on Application Accelerators in High Performance Computing* (Vol. 1, pp. 74–83). IEEE.
46. Wood, D. A., & Hill, M. D., (1995). Cost-effective parallel computing. *Computer, 28*(2), 69–72.

CHAPTER 7

RECENT DEVELOPMENTS IN CONCURRENT, PARALLEL, AND DISTRIBUTED COMPUTING

CONTENTS

7.1. INTRODUCTION

Current advancements in computer networking and architectures have made parallel, concurrent, and distributive computing systems more accessible to the general public, hence expanding the number of possible computing choices to end-users. At the extremes of the computing spectrum, two significant technological evolutions have been occurring. First, at the limited scale, processors have a growing number of independent implementation units (cores), to the point where a single CPU may be regarded as a concurrent shared-memory computer. Second, at the massive scale, the Cloud Computing technology enables applications to scale by providing resources from the massive pool on a compensation basis. To take the advantages provided by multi-core CPUs and Clouds, programs must be appropriately adapted. While being at opposite ends of the computer architecture spectrum, multi-core processors on the limited scale and Clouds on the vast scale both have one thing in common: they are both specific instances of parallel, distributed, and concurrent systems (Arroyo, 2013; Bunte et al., 2019). As a result, they bring well-known synchronization, communication, task distribution, and other issues to the developers. We examine the present status of parallel, distributed, and concurrent developments, as well as their application to multi-core architectures and cloud computing. It is found that the majority of present techniques show accessibility and adaptability issues that can impede their deployment to such new computer architectures (Chen et al., 2022; Tan et al., 2022; Villa, 1992). We proposed a modified simulation technique, depending upon the paradigm of the multi-agent system, to solve a few of these shortcomings. Although it has been doubtful that a single technique would work well in all of the aforementioned contexts, we indicate that the suggested adaptive mechanism has valuable characteristics that make it desirable in both a multi-core processor and a Cloud system. Such capabilities include the capacity to minimize communication costs by transferring simulation components and the option to add (or remove) nodes to the implementation architecture during runtime. We would also demonstrate that distributed and parallel simulations may be run on top of unstable resources with the aid of an extra support layer (Agha and Kim, 1999; Sunderam and Geist, 1999).

7.2. CURRENT DEVELOPMENTS IN CONCURRENT COMPUTING

7.2.1. Concurrency in Logic Programming

In logic programming, there are two types of parallelisms: explicit concurrency and implicit parallelism. The implicit parallelism has been primarily concerned with defining a parallel operational interpretation of Horn clause logic that is ideally comparable to the consecutive version. The objective is to increase efficiency by using a parallel computing model and a parallel architecture in the execution. Concurrency constructs are included in the pure Horn clause logic language to achieve expressive concurrency. The objective is to make logic languages more capable of dealing with concepts like concurrency and procedure. There have been a variety of concurrency models available, varying from models depending upon simple synchronization techniques to models depending upon complicated synchronization mechanisms (De Kergommeaux and Codognet, 1994; Levi, 1986; Meseguer, 1992). The necessity for clean languages to be utilized as systems programming languages in massively parallel latest generation computers is one motivator for concurrent logic languages. We'll focus on implicit parallelism inside this part, whereas concurrent logic languages would be discussed in Section 7.2.2.

7.2.2. OR-Parallelism and AND-Parallelism: Process Interpretation

The computation rule and the search rule are suggested by the computational model of logic languages as 2 features which may be addressed concurrently/ Conery81/. The usage of parallelism in the searching rule has been similar to OR-parallelism. A parallel search rule has been inherently equitable (Codish and Shapiro, 1986, 1987). The processes (which correlate to the branches in the search tree) are self-contained. Reverse communication is required: the common ancestor of a group of processes should be capable to gather the responses in a data structure that has been appropriate. The usage of parallelism in the calculation rule has been connected to AND-parallelism. The parallel assessment of genuine variable expressions in functional languages is analogous to a parallel computation rule. The primary discrepancies are due to logical parameter behavior and the lack of sequence control information. The features of AND-parallelism may be better understood in the context of 1Qgic program process evaluation.

A goal clause is understood as a network of procedures in the process of understanding (Conery and Kibler, 1985; Guala, 1999). Consider the aim in Figure 7.1, which is depicted by a network.

$\leftarrow A(x,y), B(y,w), C(x), D(x)$

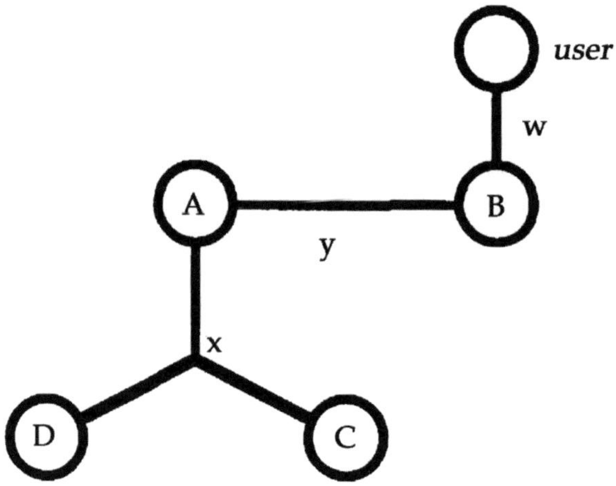

Figure 7.1. A logical process network.

Source: https://link.springer.com/book/10.1007/BFb0027037.

The network's dynamic behavior has been the procedural understanding of resolution (rewriting and unification). Unifying a procedure with a clause head correlates to output/input actions on channels, whereas rewriting involves a network reconfiguration. Thus, in a resolution, a procedure conducts output or input activities on its channels and has been replaced by the (potentially vacant) network relating to a clause body. The network's development has been nondeterministic. In theory, network procedures may be rewritten simultaneously as if they had been separated. Due to "shared" variables (channels), though, process synchronization has been required. Synchronization would be a test of the "compatibility" of the data delivered on a channel via separate processes, wherein compatible values would be those that may be unified. Synchronization may then result in failures, which have been network-wide faults (Gao et al., 2011; Kannan et al., 2021; Takemoto, 1997).

7.2.3. Variable Annotations

As the following instances demonstrate, basically read-only variables, which are indicated by the postfix function! may appear both in the objectives and in the clause bodies of an expression.

$\leftarrow +(x,2,w),+(9,w!z),*(4,w!,z)$

$*(s(x),y,z) \leftarrow *(x,y,w),+(y,w!,z)$

The identification of a parameter in a process means that process 416 must read on that channel if possible (Gupta et al., 2022; Puertas et al., 2022).

Nearly read-only parameters vary from read-only parameters in Concurrent Prolog. Even though one of the latter 2 requirements holds, a procedure is halted while using Concurrent Prolog read-only parameters. Read-only parameter annotations impact the logic program's semantics, whereas virtually read-only indications are mostly used to describe a dynamic computation rule and have no effect on the logic program's semantics. All procedures in the network have been examined for rewriting in the AND-parallel model having nearly read-only parameter annotations at every step. Such procedures that don't meet one of the aforementioned criteria are halted and may be restarted at a later stage. Annotations that are virtually read-only provide a partial ordering of the atoms in a goal. A sequential interpreter may utilize this data to determine which atom in the objective should be evaluated 1st for rewriting. After that, the programer might create his or her own dynamic calculation rule. The capability to describe a partial ordering relation increases the language's flexibility and expressiveness over Prolog, which only allows for left-to-right atomic sequencing.

$\leftarrow +(x,2,w),+(9,w!z),*(4,w!,z)$

The annotations may be seen as a technique for defining directed channels in the procedure comprehension (Figure 7.2).

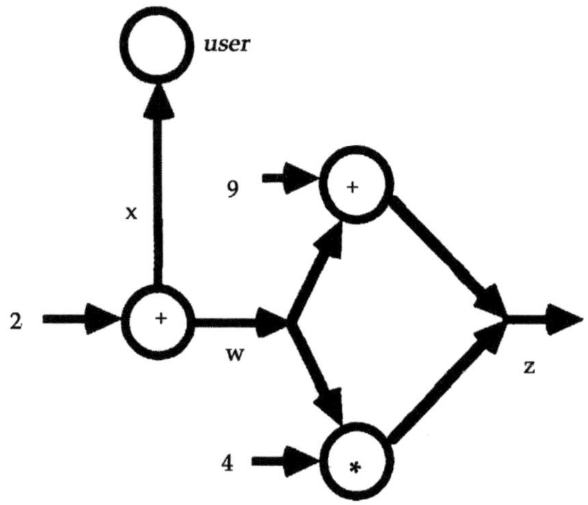

Figure 7.2. An annotated network.

Source: https://dl.acm.org/doi/book/10.5555/19518.

The channels have been redirected. All other channels have been regarded as output channels, whereas input channels correlate to annotated parameters. Generally, every channel may contain several providers and users. Constraints might be considered as channels having several providers but no customers. Unlike input and output channels, directed channels aren't purely output or input channels. In reality, procedures have been allowed to write on its "input" channels in certain instances (conditions 2 and 3). Furthermore, processes may read on their "output" channels on occasion. The network's producer-consumer interactions, on the other hand, may be productively utilized a demand-driven computing rule (parallel or serial). The outer process is the basis of the demand-driven computation rule.

7.3. RECENT TRENDS IN PARALLEL COMPUTING

In both scientific and corporate computers, computational requirements have been growing all the time. Because of the light speed as well as certain thermodynamic principles, silicon-based processor circuits have been approaching their processing speed restrictions. One way to get around this constraint is to link the many processors that are functioning in tandem with others. Parallel computing may be divided into four types: instruction-level, bit-level, data-level, and task-level parallelism. A computer effort is divided into multiple comparable subtasks that may be executed individually and whose outputs are aggregated once the job is completed in parallel computing. In the search for high-speed computing, parallel machines have become serious competitors to vector computers in recent years (D'Angelo and Marzolla, 2014; Singh et al., 2015).

7.3.1. Hardware Architectures for Parallel Processing

CPUs are the key components of parallel computing. Based on commands and data streams that may be concurrently processed, computer systems may be categorized as follows:

- Single instruction single data (SISD);
- Single instruction multiple data (SIMD);
- Multiple instruction single data (MISD);
- Multiple instruction multiple data (MIMD).

7.3.1.1. Single Instruction Single Data (SISD)

A single uni-core processor runs a single instruction stream to act on information kept in a separate memory in a SISD computer system. The SISD model has been used in the majority of conventional computers. The primary memory must hold all of the commands and information. The pace at which the computer may alter the data internally determines the processor element's speed. Macintosh, Workstations, and other significant instances of the paradigm may be found. Among the most recent SISD computers, superscalar processors and pipelined processors are prominent examples (Figure 7.3) (Ibaroudene, 2008; Quinn, 2004; Singh et al., n.d.).

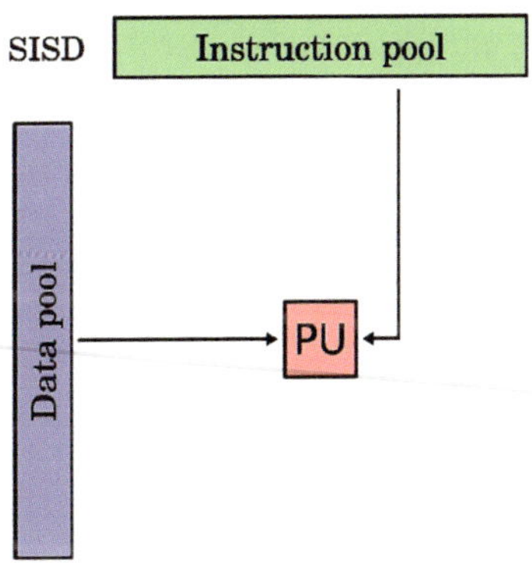

Figure 7.3. Single instruction single data (SISD).

Source: https://commons.wikimedia.org/wiki/File:SISD.svg.

7.3.1.2. Single Instruction Multiple Data (SIMD)

Single instruction multiple data (SIMD) refer to a multiprocessor system that can run the same command on all CPUs while using distinct data streams. Because they include a lot of matrix and vector operations, machines depending upon this approach are ideally suited for scientific computing (Liao and Roberts, 2002).

7.3.1.3. Multiple Instruction Single Data (MISD)

The MISD computer system is a multiprocessor machinc that can execute various commands on separate processing components while all working on the same data set. The majority of applications don't benefit from machines that adopt this paradigm (Flynn, 1966; Flynn and Rudd, 1996).

7.3.1.4. Multiple Instruction Multiple Data (MIMD)

A multiprocessor computer able to perform multiple commands on multiple data sets has been the MIMD computing system. Because every processing component in the MIMD model has its command and data stream, machines designed with this paradigm are suitable for a variety of applications (Azeem, Tariq, and Mirza, 2015; Hord, 2018; Pirsch and Jeschke, 1991). Depending on how many of the processing parts have been connected to the main memory, they are divided into distributed memory MIMD and shared memory MIMD machines. The concurrent computational model may be classified into three groups based on the memory model:

- Shared memory computational model;
- Distributed computational model; and
- Hierarchical memory model.

7.3.2. Shared Memory Parallel Computing Models

A shared memory architecture connects a group of processors to a centralized memory. Data exchange is quick because all processors share a single address space, yet processes may damage every other's data at the same time. To protect the data from corruption, locks, and semaphores have been utilized. Processing components and memory have restricted scalability, which implies we can't add as many processing components because we need a finite memory (Figure 7.4).

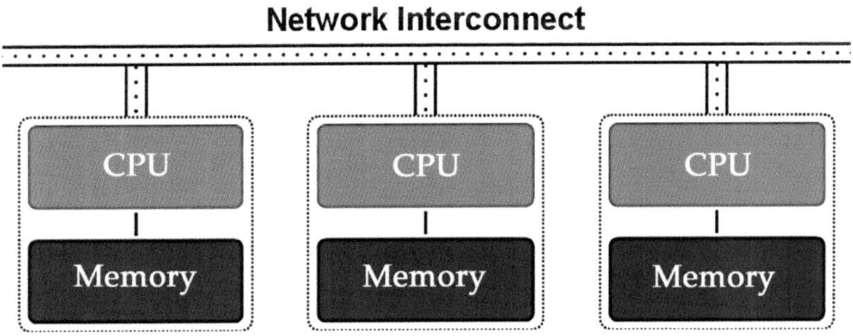

Figure 7.4. Shared memory parallel computing models.

Source: https://stackoverflow.com/questions/53578654/shared-memory-model-and-distributed-memory-model.

7.3.2.1. PRAM Model

Wyllie, Fortune, and Goldschlager created the PRAM model, which has been commonly utilized shared memory computing model for the construction and study of parallel algorithms. A limited number of processors share a global memory pool. The processors have been permitted to view memory simultaneously, and each operation takes a single unit of time to perform. There are several examples of the PRAM model. The CRCW PRAM paradigm (Joseph, 1992) allows for write and read operations on the same memory cell at the same time. Another form, the CREW PRAM (Joseph, 1992), allows several processors to read from a similar memory cell at the same time, but only one CPU may write to a cell at a time. The PRAM concept is simple to construct, however, it has network and memory congestion issues.

7.3.2.2. Distributed Memory Parallel Computing Models

There have been a variety of parallel computing architectures with distributed memory whereby every machine with its memory has been linked to a communication system. LogP models and BSP Models belong to this group. Such approaches eliminate the limitations of shared memory computing memory (Gregor and Lumsdaine, 2005; Killough and Bhogeswara, 1991).

7.4. RECENT TRENDS IN DISTRIBUTED COMPUTING

Only a few years ago, to receive data and utilize resources, a user needed infrastructure and a computer system in the same physical area, like data storage devices. Grid computing emerged in the mid-1990s intending to allow users to remotely access optimal processing capacity in other computing centers whenever local computing centers are overloaded (Curnow and Wichmann, 1976). Subsequently, researchers devised the concept of providing not just on-demand computing power, and also software, storage devices, and data. Eventually, cloud computing emerged to provide 3rd parties with dependable, adaptable, and low-cost resources (Mittal, Kesswani, and Goswami, 2013; Salvadores and Herero, 2007). Cloud computing was introduced at the end of 2007, but the introduction of cloud computing eliminated the requirement for infrastructure in the same area. We have access to kcy data from any location at any timc. Virtualization has been the fundamental component of Cloud Computing. It essentially delivers a paid resource on demand. Only an internet connection has been required to access cloud services.

7.4.1. Cluster Computing

A cluster is a collection of distributed or parallel processing systems containing a large number of desktop computers that work together to form a regular high computer system. Clusters have been primarily utilized for highly available, balancing of load, and processing power (Baker, Fox, and Yau, 1995; Sadashiv and Kumar, 2011).

Because certain issues in business, engineering, and science might not be addressed by supercomputers, they had been tried in clusters to overcome the issues of higher computation and the expense of addressing them using supercomputers (Curnow and Wichmann, 1976). A cluster's elements communicate with one another over a rapid local area network (LAN). Several computers have been linked in a cluster and function like a single virtual machine that shares its task in a cluster architecture. Whenever a user works, the server receives the request and sends it to all of the independent computers for processing. As Consequently, workloads are distributed over thousands of isolated machines, allowing for quicker processing and greater computing. Cluster, on the other hand, is divided into three pieces. Higher-availability clusters, load-balancing clusters, and HPC clusters are all examples of high-availability clusters. Clusters have the advantages of availability, scalability, and higher efficiency (Baker and Buyya, 1999; Sadashiv and Kumar, 2011).

Petroleum reservoir modeling, google search engine, protein explorer, image rendering, and earthquake simulation are all major cluster applications.

Elasticity, scalability, middleware, software, encryption, and security, balancing of load, and management have all been major concerns in cluster computing. Clusters were created primarily for database applications. For instance, in Windows, an active directory cluster may be used, and in web development, a My SQL database cluster may be used (Figure 7.5).

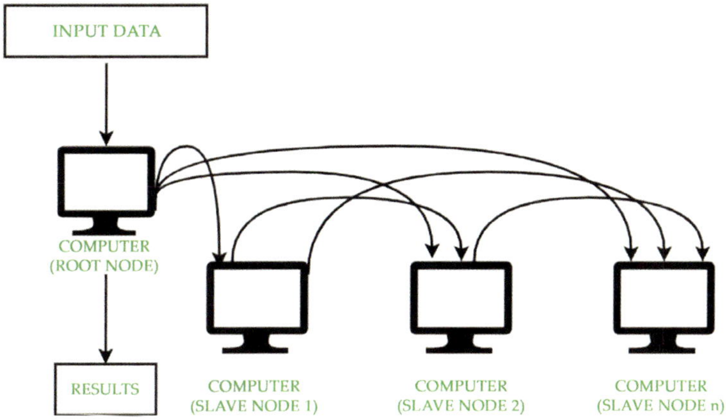

Figure 7.5. Cluster computing architecture.

Source: https://www.geeksforgeeks.org/an-overview-of-cluster-computing/.

As seen in Figure 7.5, there are 4 boxes A, B, C, and D which are all linked to one another. A cluster's characteristic has been that it replicates data. As a result, each computer in a cluster has the same data as the others. Each machine has an OS and a SQL database installed on it. Assume a consumer accesses the database on computer D via the internet. Since computer D already has sufficient applications, it would be instantly sent to computers A, B, or C due to replication. The cluster has been based on this principle. Assume, however, that computer B has failed, as represented by E in the diagram. Whenever computer B fails, A would immediately disconnect from B and link to C, and whenever computer B comes back up (in this case, F), it would reconnect to B.

7.4.2. Grid Computing

Grid computing may simply be defined as the next phase in the distributed computing process. The electrical grid, sometimes known as the power grid, inspired grid computing. The electrical grid's wall outlets assist in connecting to resources, as well as generating and distributing power bills. When a client authenticates to the power grid, he doesn't have to know where the power plant is physically located. All a client cares about are having the required power for his needs. Grid computing follows the same principle; it employs a middleware layer to coordinate various IT resources over a network that functions as a virtual whole (Figure 7.6).

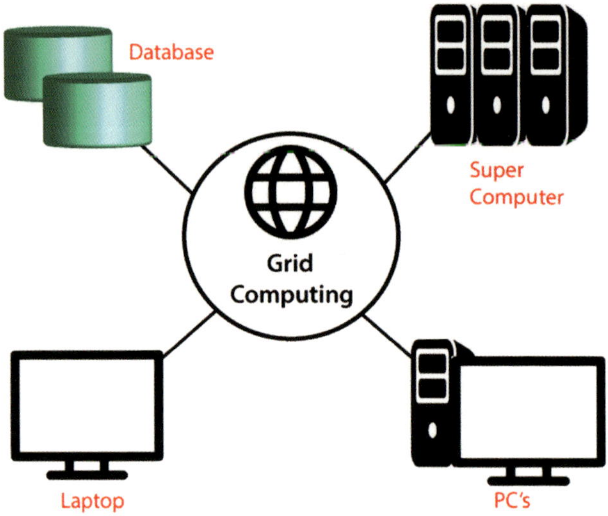

Figure 7.6. Grid computing.

Source: https://www.tutorialandexample.com/grid-computing-vs-cloud-computing.

Grid computing aims to generate the appearance of huge and powerful self-managing virtual systems by connecting a large number of heterogeneous computers that share a variety of resources (Joseph, 1992).

To interface with the diverse data set and hardware, it employs a layer of middleware. Grid computing is a type of computation that has been built on distributed and parallel computing concepts and internet protocols. It allows users to share computational sources, storage components, specialized programs, and equipment, among other things (Dongarra, 1993). Grid computing has 2 basic goals: one of which is to reduce overall calculation time while using current technology, and the other is to make the most of idle computer and hardware sources (Weicker, 1984). Grid employs virtual groups to share geographically dispersed sources for achieving a shared aim. Virtual organizations are described in grid computing as "a collection of individuals or institutions that share the computer sources of a grid for a shared aim" (Jadeja and Modi, 2012). LHC, SETI@home, climateprediction.net, and Screensaver lifesaver are among the most extensively utilized apps on personal computers. Elevated-energy physics, Computational chemistry, Earth science, biology, Fusion, Astrophysics, and other areas are examples of grid computing. Scalability, h Heterogeneity, adaptive or fault tolerance, and security have been some of the major attributes given by computation grids. The middleware, consumer, and resource levels are the three key elements that make up a grid.

7.4.3. NKN Project

The "National Knowledge Network" is a project that has been recognized by the State of India. The purpose of such a project is to create a 10 Gbps data communication network that will link all colleges and institutes of higher learning and research to encourage information exchange and collaborative research. It now has seven super nodes and 24 core nodes in operation, connecting 628 institutions, with the remaining 850 expected to be linked by March 2013. The Government of India started this initiative, which is a grid computing effort that uses optimal data and resources from universities. This is the most significant step forward in e-learning, with various applications in education, agriculture, e-governance, and health.

7.4.4. Cloud Computing

Cloud computing is a progression of grid computing that allows us to obtain more or expand infrastructure-based services. The cloud is a novel service that allows consumers to access resources such as storage and processing on-demand (Ferreira et al., 2003, n.d.).

Cloud computing offers massively scalable computer sources as a pay-per-use service. One of the most important applications of cloud computing has been that the user just utilizes the services which he requires. We don't have to bother about how resources or servers have been managed behind the scenes; instead, we just lease or rent resources as needed. Utility computing or 'IT on demand' are other terms for cloud computing. It's a brand-new business model that takes advantage of cutting-edge IT technologies such as multi-tenancy and virtualization. Each of these services is utilized to benefit from economies of scale and lower IT resource costs. Google applications and Microsoft SharePoint (Krishnadas, 2011) are two common examples of cloud services (Figure 7.7).

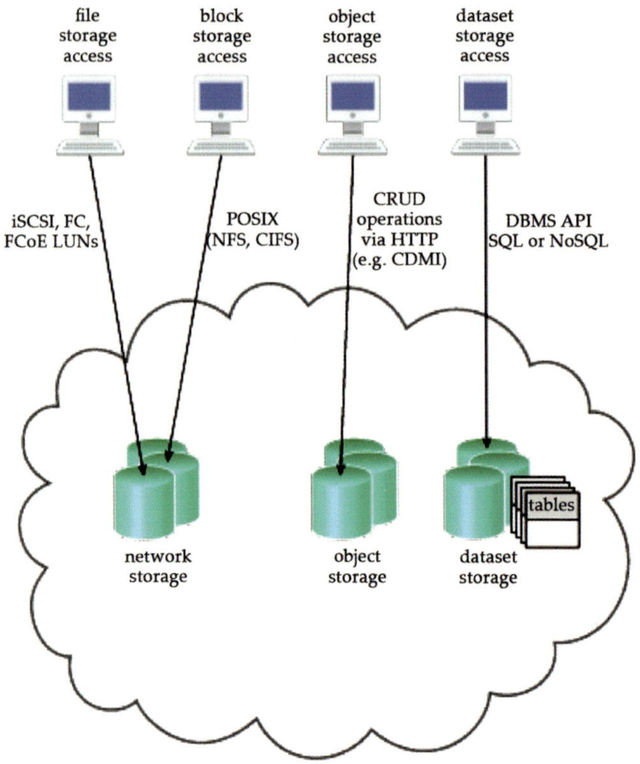

Figure 7.7. Cloud computing mechanism.

Source: https://patterns.arcitura.com/cloud-computing-patterns/mechanisms/cloud_storage_device.

Here is an example of how cloud computing may be used in the real world. The usage of an exchange server for a private mailing service has been demonstrated by the following firm, as shown in the figure. Imagine that this company has 200 hosts and that all of them use the services provided by the exchange server. When a server goes offline or crashes, all of the company's work would come to a halt, which would result in a significant loss of both resources and money. We rely on cloud services to eliminate this kind of catastrophic event (Androcec, Vrcek, and Seva, 2012; Chen, Chuang, and Nakatani, 2016; Younas, Jawawi, Ghani, Fries, and Kazmi, 2018).

As evident from the diagram, the company's exchange server has been hosted in the cloud through a cloud service provider and has been paid for on a pay-per-use basis. Therefore, even when a company's server crashes or goes offline, the business should not be concerned since the server would always be kept online by the cloud's backup services. It is one of the cloud computing industry's most essential building blocks. The following are the four primary categories of cloud computing:

- Private cloud;
- Public cloud;
- Hybrid cloud; and
- Community cloud.

The previous illustration shows a private cloud. A cloud that is used by 2 or more companies simultaneously is referred to as a community cloud, while a cloud that is accessible by anyone with an internet connection is known as a public cloud. Reliability, agility, elasticity, quality of service, adaptability, and virtualization are some of the other important characteristics of cloud computing. Reduced operating expenses and improved user experience are two advantages of cloud computing (Doshi and Kute, 2020).

REFERENCES

1. Agha, G. A., & Kim, W., (1999). Actors: A unifying model for parallel and distributed computing. *Journal of Systems Architecture, 45*(15), 1263–1277.
2. Androcec, D., Vrcek, N., & Seva, J., (2012). *Cloud Computing Ontologies: A Systematic Review.* Paper presented at the proceedings of the third international conference on models and ontology-based design of protocols, architectures and services.
3. Arroyo, M., (2013). *Teaching Parallel and Distributed Computing to Undergraduate Computer Science Students.* Paper presented at the 2013 IEEE International Symposium on Parallel & Distributed Processing, Workshops and PhD Forum.
4. Azeem, M., Tariq, A., & Mirza, A., (2015). A review on multiple *Instruction Multiple Data (MIMD) Architecture.* Paper presented at the international multidisciplinary conference.
5. Baker, M., & Buyya, R., (1999). Cluster computing at a glance. *High Performance Cluster Computing: Architectures and Systems, 1*(3–47), 12.
6. Baker, M., Fox, G. C., & Yau, H. W., (1995). *Cluster Computing Review.*(pp. 2-9)
7. Bunte, O., Groote, J. F., Keiren, J. J., Laveaux, M., Neele, T., Vink, E. P. D., & Willemse, T. A., (2019). *The mCRL2 Toolset for Analyzing Concurrent Systems.* Paper presented at the international conference on tools and algorithms for the construction and analysis of systems.
8. Chen, T., Chuang, T. T., & Nakatani, K., (2016). The perceived business benefit of cloud computing: An exploratory study. *Journal of International Technology and Information Management, 25*(4), 7.
9. Chen, Y., Liu, B., Cai, L., Le Boudec, J. Y., Rio, M., & Hui, P., (2022). Guest editorial: New network architectures, protocols, and algorithms for time-sensitive applications. *IEEE Network, 36*(2), 6–7.
10. Codish, M., & Shapiro, E., (1986). *Compiling OR-Parallelism into AND-Parallelism.* Paper presented at the International Conference on Logic Programming.
11. Codish, M., & Shapiro, E., (1987). Compiling OR-parallelism into AND-parallelism. *New Generation Computing, 5*(1), 45–61.
12. Conery, J. S., & Kibler, D. F., (1985). AND parallelism and

nondeterminism in logic programs. *New Generation Computing, 3*(1), 43–70.

13. Curnow, H. J., & Wichmann, B. A., (1976). A synthetic benchmark. *The Computer Journal, 19*(1), 43–49.
14. D'Angelo, G., & Marzolla, M., (2014). New trends in parallel and distributed simulation: From many-cores to cloud computing. *Simulation Modelling Practice and Theory, 49*, 320–335.
15. De Kergommeaux, J. C., & Codognet, P., (1994). Parallel logic programming systems. *ACM Computing Surveys (CSUR), 26*(3), 295–336.
16. Dongarra, J. J., (1993). *Performance of Various Computers Using Standard Linear Equations Software*. University of Tennessee, Computer Science Department.
17. Doshi, R., & Kute, V., (2020). *A Review Paper on Security Concerns in Cloud Computing and Proposed Security Models*. Paper presented at the 2020 International Conference on Emerging Trends in Information Technology and Engineering (IC-ETITE).
18. Ferreira, L., Berstis, V., Armstrong, J., Kendzierski, M., & Neukoetter, A., (2003). In: Masanobu, T., et al., (eds.), *Introduction to Grid Computing with Globus*. IBM Redbooks.
19. Ferreira, L., Berstis, V., Armstrong, J., Kendzierski, M., Neukoetter, A., Takagi, M., Murakawa, R., Olegario, H., James, M., & Norbert, B.(2003) *Introduction to Grid Computing with Globus*.
20. Flynn, M. J., & Rudd, K. W., (1996). Parallel architectures. *ACM Computing Surveys (CSUR), 28*(1), 67–70.
21. Flynn, M. J., (1966). Very high-speed computing systems. *Proceedings of the IEEE, 54*(12), 1901–1909.
22. Gao, L., Yang, J., Zhang, H., Zhang, B., & Qin, D., (2011). FlowInfra: A fault-resilient scalable infrastructure for network-wide flow level measurement. *Paper Presented at the 2011 13th Asia-Pacific Network Operations and Management Symposium.*
23. Gregor, D., & Lumsdaine, A., (2005). Lifting sequential graph algorithms for distributed-memory parallel computation. *ACM SIGPLAN Notices, 40*(10), 423–437.
24. Guala, F., (1999). The problem of external validity (or "parallelism") in experimental economics. *Social Science Information, 38*(4), 555–573.
25. Gupta, R., BV, S., Kapoor, N., Jaiswal, R., Nangi, S. R., & Kulkarni,

K., (2022). *User-Guided Variable Rate Learned Image Compression.* Paper presented at the proceedings of the IEEE/CVF conference on computer vision and pattern recognition.

26. Hord, R. M., (2018). *Parallel Supercomputing in MIMD Architectures*: CRC press.
27. Ibaroudene, D., (2008). *Chapter 1: Motivation and History, Parallel Processing*. St. Mary's University, San Antonio, TX. Spring.
28. Jadeja, Y., & Modi, K., (2012). *Cloud Computing-Concepts, Architecture and Challenges.* Paper Presented at the 2012 international conference on computing, electronics and electrical technologies (ICCEET).
29. Joseph, J., (1992). *An Introduction to Parallel Algorithms*. In: Addison-Wesley.
30. Kannan, P. G., Budhdev, N., Joshi, R., & Chan, M. C., (2021). *Debugging Transient Faults in Data Centers Using Synchronized Network-Wide Packet Histories.* Paper presented at the 18th USENIX Symposium on Networked Systems Design and Implementation (NSDI 21).
31. Killough, J. E., & Bhogeswara, R., (1991). Simulation of compositional reservoir phenomena on a distributed-memory parallel computer. *Journal of Petroleum Technology, 43*(11), 1368–1374.
32. Krishnadas, N., (2011). *Designing a Grid Computing Architecture: A Case Study of Green Computing Implementation Using SAS®.* Paper presented at the SAS Global Forum.
33. Levi, G., (1986). Logic programming: The foundations, the approach and the role of concurrency. *Current Trends in Concurrency*, 396–441.
34. Liao, Y., & Roberts, D. B., (2002). A high-performance and low-power 32-bit multiply-accumulate unit with single-instruction-multiple-data (SIMD) feature. *IEEE Journal of Solid-State Circuits, 37*(7), 926–931.
35. Meseguer, J., (1992). Conditional rewriting logic as a unified model of concurrency. *Theoretical Computer Science, 96*(1), 73–155.
36. Mittal, G., Kesswani, D., & Goswami, K., (2013). *A Survey of Current Trends in Distributed, Grid and Cloud Computing.* arXiv preprint arXiv:1308.1806.
37. Pirsch, P., & Jeschke, H., (1991). *MIMD (Multiple Instruction Multiple Data) Multiprocessor System for Real-Time Image Processing.* Paper presented at the image processing algorithms and techniques II.

38. Puertas, E., De-Las-Heras, G., Sánchez-Soriano, J., & Fernández-Andrés, J., (2022). Dataset: Variable message signal annotated images for object detection. *Data, 7*(4), 41.
39. Quinn, M. J., (2004). *Parallel Programming in C with MPI and OpenMP (in Chinese), Chen WG, Wu YW Trans.* Beijing: Tsinghua University.
40. Sadashiv, N., & Kumar, S. D., (2011). *Cluster, Grid and Cloud Computing: A Detailed Comparison.* Paper presented at the 2011 6th international conference on computer science & education (ICCSE).
41. Salvadores, M., & Herero, P., (2007). *Virtual Organization Management over Collaborative Awareness Model (CAM).* (Vol. 1, pp.1-12).
42. Singh, U. L., Mahato, D. P., Singh, R. S., & Khosrow-Pour, M (2015). Recent trends in parallel computing. *Encyclopedia of Information Science and Technology*, 3580–3589.
43. Sunderam, V. S., & Geist, G., (1999). Heterogeneous parallel and distributed computing. *Parallel Computing, 25*(13, 14), 1699–1721.
44. Takemoto, M., (1997). *Fault-Tolerant Object on Network-Wide Distributed Object-Oriented Systems for Future Telecommunications Applications.* Paper presented at the proceedings Pacific rim international symposium on fault-tolerant systems.
45. Tan, N., Bird, R. F., Chen, G., Luedtke, S. V., Albright, B. J., & Taufer, M., (2022). Analysis of vector particle-in-cell (VPIC) memory usage optimizations on cutting-edge computer architectures. *Journal of Computational Science, 60*, 101566.
46. Umrao, L. S., Mahato, D. P., & Singh, R. S., (2015). Recent trends in parallel computing. In: *Encyclopedia of Information Science and Technology* (3rd edn., pp. 3580–3589). IGI Global.
47. Villa, F., (1992). New computer architectures as tools for ecological thought. *Trends in Ecology & Evolution, 7*(6), 179–183.
48. Weicker, R. P., (1984). Dhrystone: A synthetic systems programming benchmark. *Communications of the ACM, 27*(10), 1013–1030.
49. Younas, M., Jawawi, D. N., Ghani, I., Fries, T., & Kazmi, R., (2018). Agile development in the cloud computing environment: A systematic review. *Information and Software Technology, 103*, 142–158.

CHAPTER 8

FUTURE OF CONCURRENT, PARALLEL, AND DISTRIBUTED COMPUTING

CONTENTS

8.1. INTRODUCTION

Concurrent, parallel, and distributed computing has been crucial to technological development over the last 30 years because it combines computational power and an effective and potent hardware system (Li, Jiao, He, and Yu, 2022; Mampage, Karunasekera, and Buyya, 2022).

Throughout the years, concurrent, parallel, and distributed systems have taken over the traditional supercomputing resources, such as science-based modeling. The modern systems now span into artificial intelligence (AI), machine learning (ML), and large-scale communications systems, such as social networks or transport systems. Therefore, this rapidly emerging field must solve a lot of present and upcoming obstacles to fulfill the fast-expanding computing demands of new and bigger highly distributed programs (Zarei, Safari, Ahmadi, and Mardukhi, 2022; Zomaya, 1996).

8.2. FUTURE TRENDS IN CONCURRENT COMPUTING

In Dutch science and research management for computer sciences, listing topics, topics, or areas of concentration is becoming popular. Frequently, these lists give renames for what individuals did before, and despite a language change, we all continue performing the same things. (For example, several formal approaches would be renamed "the computer of the future.") The basic purpose of such lists is to offer Dutch computer science and research a fresh emphasis. That paper included a listing of topics, but just for the sake of speculating on what individuals may do in the next years. Under no way can these topics be deemed a concern for their own sake simply because they are included here. Intriguingly, this dispassionate method renders the study less 'scientific' since predicting human behavior is not part of computer science, although regulating their (research) behavior may be. In any case, we do not aim to change themes for no reason (Li et al., 2022; Sunderam, 1990; Zomaya, 1996).

8.2.1. Dynamics and Interaction

Concurrency is the behavior of systems that communicate. It provides methods for describing dynamics and interaction about such representations. It is now regarded as a branch of computer science study and is used in systems that are applied in software or hardware. We regard a breakthrough to be significant if it has applications totally beyond the area of computer science. It is most prevalent in the biological sciences, where parallelism concepts are utilized to understand the behavior and interactions inside a live cell. But parallelism theories can also be used in mechatronics and systems engineering to describe the behavior and connection of machinery, assembly plants, and automated industries. This is evidence of the development of concurrency theories.

8.2.2. Grid Computing and Cloud Computing

Grid computing and cloud computing are undeniably significant trends. Both advancements rely substantially on the creation of unique protocols. Because security-based protocols are ubiquitous, their complexity is remarkable. Existing concurrent theory is too basic to explore this new territory, yet if we have trust in the efficacy of such approaches, they would appear as effective instruments in the study of the aforementioned protocols. There is a good chance that it would be the focus of initiatives around the globe (Figure 8.1) (Foster, Zhao, Raicu, and Lu, 2008; Gill et al., 2022; Murad et al., 2022; Saini, Kalra, and Sood, 2022; Yousif et al., 2022).

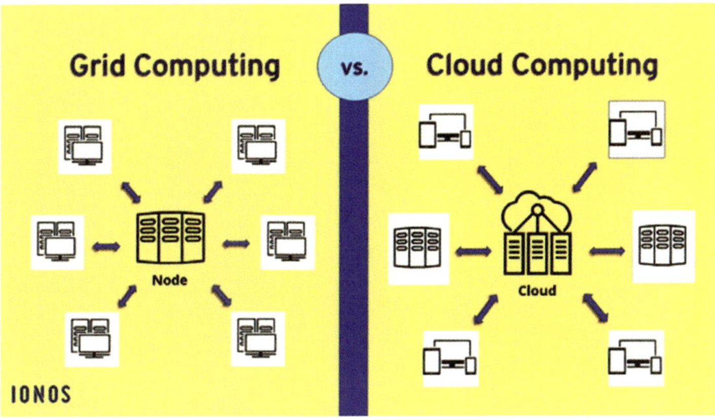

Figure 8.1. Grid computing and cloud computing.

Source: https://www.ionos.com/digitalguide/server/know-how/grid-computing/.

8.2.3. Hybrid Systems

Historically, concurrency theory analyzes the kinetics as well as the interconnection of discrete-event systems. Embedded system applications in particular require the modeling of continually developing physical beings, which are often represented by differential equations. In various methods, discrete-event formalisms are supplemented with differential (algebraic) equations to represent and analyze these hybrid phenomena. The most effective model of a hybrid system is a hybrid automaton (Cimatti, Mover, and Tonetta, 2011; Tan, Kim, Sokolsky, and Lee, 2004; Xu and Wu, 2022). Such hybrid automata already exist in a variety of flavors with corresponding testing tools. There are other hybrid procedure algebras. The difficulty is in establishing a link with the dynamics and management field, which includes models like spline affine networks, mixed logic complex systems, and linear complementing systems. In dynamics and control, regulator design and study of features like steadiness and testability are emphasized (Figure 8.2).

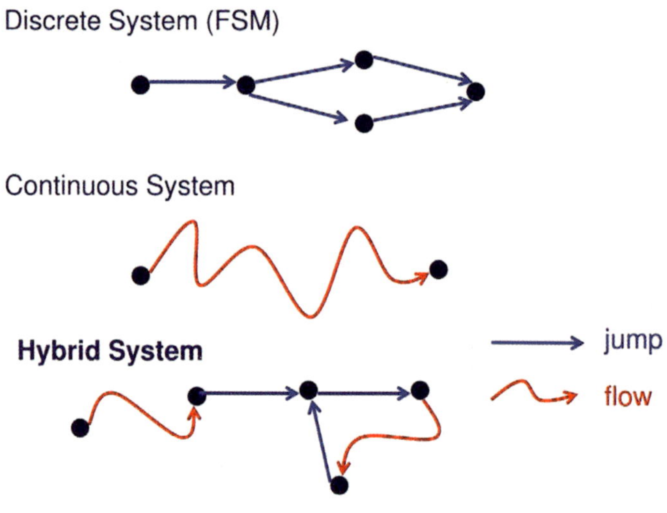

Figure 8.2. Hybrid systems.

Source: https://www.slideserve.com/aliza/hybrid-systems-concurrent-models-of-computation.

8.2.4. Mobility and Security

Continued research on movement as well as calculi to convey the movement of systems. With the examination of security procedures, one of the most significant application areas is security. Concerning safety, we anticipate that security components will be included in all other modes as well as protocols of interaction. In this respect, security will stop being a distinct topic, much like efficiency is an issue that must constantly be considered.

Incorporation of security elements into established functions as well as their formal definitions would need a substantial amount of future effort. The majority of this research would be conducted within the framework of a few of the (somewhat algebraic) theories (Pigatto et al., 2014; Ravi, Raghunathan, Kocher, and Hattangady, 2004; Skoglund, 2022; Sutar and Mekala, 2022).

8.2.5. Agents and Games

Apart from concurrency theory, the notion of AI techniques has emerged. We believe that concurrency theory principles will be used in the continuous work of agency theory. Moreover, ties to game theory are essential. Activities in gaming might result in agent programming notations dependent on concurrency theory or heavily impacted by it. Ultimately, gaming agents would be engaged in incredibly complex communication systems that would reflect creatures that may seem realistic but do not have to adhere to the rules of physics. In a game, time travel is permissible, but a theory of temporal concurrency admits that such advances might well be elusive. Many other theories would work their way into the development of agents in gaming, but for the communication aspect of things, concurrency theories might offer undiscovered potential when the rising intricacy of massively networked games causes significant problems of this sort.

8.2.6. Natural Computing

Concurrency theory would be linked to ideas from several fields of natural computing, including quantum computing as well as relativistic computing in spacetime. Quantum computing is useful for modeling parallelism on a microchip, as well as classical characteristics are useful for simulating interaction with long delays, such as that experienced in space flight (de Castro, 2007).

8.2.7. Caveat

A modest overview of future paths we published in 1996 was well received by some. We hypothesize on a few different possibilities once again. Some have wholly distinct perspectives on future trends, and our pick of six topics may probably overlook a movement that would be significant in the next years.

8.3. FUTURE TRENDS IN PARALLEL COMPUTING

The demand for computing across all areas of science, engineering, entertainment, government, and despite the rising technological difficulties in allowing such expansion, the need for computers in all sectors of research, engineering, entertainment, government, as well as business products would keep expanding. As empirical and analytical information is merged with simulation, computer applications would grow increasingly complicated and dispersed, blurring the distinctions between various sorts of applications as well as systems. Engineers and scientists previously dominated parallel computing, which has been used to model physical processes ranging from biological macromolecules and fluid dynamics to earth's environment and astrophysics. However, with the growth of online, clinical, economic, as well as genetic code sources, parallelism is increasingly required across all computer areas, not only in the physical sciences. Deep neural networks are among the most computational complexity applications on present systems, and they frequently work on huge volumes of data or explore large state spaces, which would increase the necessity for higher speed and parallelism. The supply for more expense, power, as well as competent computing systems would then develop all over basic science, technology, company, government, as well as entertainment, driven by current societal important targets of recognizing highly complicated processes in healthy and natural systems. Severe parallel, as well as distributed computing, is required for three types of uses which are discussed in subsections.

8.3.1. Machine Learning (ML)

Several applications in science, industry, government, as well as other sectors, are being revolutionized by ML's ability in deriving patterns from big data sets, either created or collected. ML applications would keep growing as more data is gathered, filtered, and shared across more domains (Figure 8.3) (Mitchell and Mitchell, 1997; Sarker, 2021).

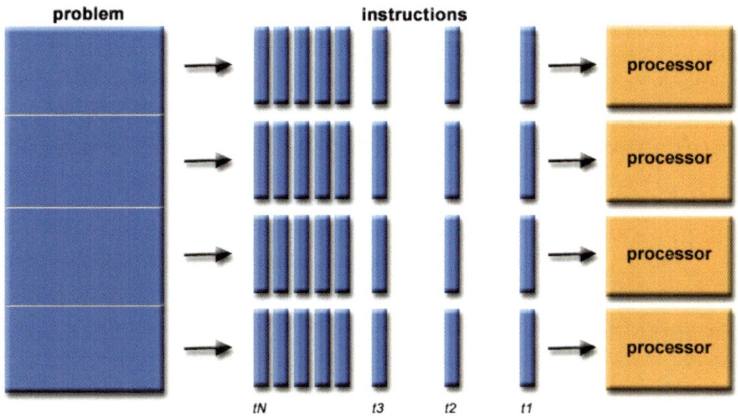

Figure 8.3. Parallel problem overview.

Source: https://www.kdnuggets.com/2016/11/parallelism-machine-learning-gpu-cuda-threading.html.

ML techniques will be able to function on dispersed data that is not completely shared and might be encoded, driving computing needs to the edge, thanks to new federated learning approaches. In the past eight years, the quantity of data, as well as compute necessary to learn the most effective picture classification problems, has increased threefold.

8.3.2. Data Analysis

Data sets keep expanding in size, owing to improved resolution, speed, and cheaper costs of cameras, detectors, sequencers, and other devices. At the same time, there is an ever-increasing connection among people and organizations, making data collection and integration simpler. Access to data has also been aided by government laws encouraging an open exchange of federally supported scientific initiatives. Whether putting adverts or investigating microbial relationships in the environment, the advantages of merging data across modes are widely established. Existing practices addressing the opioid addiction issue, for example, involve establishing biosignatures for dependence spanning-omics (genomics, etc.), outcome, sensor, as well as health record data for huge populations, including analytic streams possibly reaching into the zettabytes (Dartnell, Desorgher, Ward, and Coates, 2007; Mansor et al., 2018; Patty et al., 2019).

8.3.3. Simulation

High-throughput simulation projects to produce enormous arrays of simulated findings, like the Materials Genome Initiative as well as weather prediction research, have broadened scientific simulation above standard HPC simulations. To assess proposed environmentally sustainable strategies, severe multi-faceted models of whole ecosystems are required. In applications like as diagnostic imaging, massive scientific studies, and dispersed sensing in cell devices, simulations could be used with efficiency to tackle inverse issues. Reverse problems, that try to determine the causative elements that created a dataset from a sequence of observations, raise processing demands by hundreds or even thousands of times above today's greatest simulations but provide new adaptive avenues to research and risk measurement. Inverse issues are utilized in big experimental equipment to rebuild 3D microstructures from x-ray imaging, interpret MRI information from clinical testing, and increase our knowledge of the basic rules of physics by comparing models to cosmological or other facts (Asch et al., 2018; Matouš, Geers, Kouznetsova, and Gillman, 2017; Wilkins-Diehr, Gannon, Klimeck, Oster, and Pamidighantam, 2008).

Such applications are not different, but they will become more confused and mixed in the future. Specialists in the social sciences might need natural language to communicate with automated development tools that squeeze information from multiple sites on the web, construct predictive methods of complicated social behavior, and then use models to forecast demographic shifts and evolving needs for housing, medical services, and other infrastructural facilities. Urban areas would be equipped with sensors that monitor energy usage in buildings and thus will control energy generation, storage, and distribution via a sophisticated electricity network of dynamic alternative energy sources. Urban areas will also maximize ease, affordability, and fuel efficiency by using efficient transport predictions made from several data sources. Self-driving cars would collect and interpret certain data in real-time, but they will also exchange information with surrounding vehicles and computers. Data from vibrational sensors, biosensors, carbon sensors, and other sensors will be mined to better understand earthquake safety and environmental concerns, as well as to help in the construction of safe structures, pollution control, and digital agriculture. The massive task of feeding the world's rising population will be addressed through digital agriculture. It will be heavily data-driven, requiring the integration of heterogeneous data from a variety of in-ground or on-ground sensors (electrochemical, mechanical, airflow, and so on), as well as sensors in drones, satellites, and other devices. It will regulate planting, watering, fertilizing, and harvesting schedules using ML and some other data processing, and (in certain situations) conduct those tasks robotically. Such a system has to be able to forecast regular activity, responding to variations in the time of day, season, and societal routines of usage, as well as severe occurrences such as natural catastrophes, cyberattacks, physical assaults, and system failures (Fox and Glass, 2000; Schultz, Rosenow, and Olive, 2022; Wu et al., 2021).

8.3.4. Beyond Digital Computing

With the termination of Moore's Law, there are fresh chances to give insight to computing platforms that might result in orders of magnitude increases in accessible computing resources by shattering the digital electronic assumption, in which calculations are assumed to output specific values. Emerging computer models, like quantum computing, stochastic computing, as well as neuromorphic computing, assure severe calculations while adhering to modern computing restrictions (e.g., energy or power), with the caveat that these computational methods could only calculate outcomes to a certain accuracy or probability. Whereas all of these systems want to offer great possibilities if accepted, embracing estimated computations (such as developing new abstract ideas to explain and modify them) opens the door to new numerical methods that introduce fresh adaptability to current systems, like enhanced parallelism by enabling direct racing events with appropriate bounded likelihood (Lammie, Eshraghian, Lu, and Azghadi, 2021; Navau and Sort, 2021; Schuld, Sinayskiy, and Petruccione, 2014).

8.3.5. Challenges in Parallel Computing

Abstractions might well be created with several objectives in mind, such as efficiency, stability, accessibility, versatility, compositionality, reusability, and verification. A system that enables people to control the specifics of moving data, energy usage, and resilience could be challenging to use, while a system that extracts such factors from users as well as handles them instantly may incur substantial time or memory requirements and may not handle those well for other apps. All of these needs may be evaluated within a specific use environment, such as energy requirements, system size, or application demand. It requires more study on how to build concepts for parallel/distributed systems that achieve these objectives, to what degree they may be concealed, as well as though they must be exposed in an attractive and useful interface (Lorenzon and Beck Filho, 2019).

The conflict between universal abstractions as well as domain-specific representations is also eternal. Power control, for instance, can be performed in a generic manner depending on the overall load demand, or it can be performed using an application catalog that requires an account of individual data types and computational methods, e.g., distributing frames in such a responsive mesh generation structure can maximize for both task scheduling and locality, but this software tool is not recyclable in other areas. Consider critical examples in the parallel/distributed design space, such as physical simulations to bounteous data parallelism utilized on batch-scheduled HPC equipment and innately dispersed IoT applications that use ML on a pair of heterogeneous computing endpoints that contains both tiny devices as well as cloud data endpoints (Hwu, 2014; Sadiku, Musa, and Momoh, 2014).

8.3.6. Risks in Parallel Computing

Genuine users, applications, and workloads covering a variety of application areas should influence the creation of abstractions to guarantee that they satisfy real demands. On the other side, we would attempt to build abstractions that are relevant to a wide range of applications and are not overly particular to a few (Tchernykh, Schwiegelsohn, Talbi, and Babenko, 2019; Zeng, 2022). The development of benchmark sets that represent a sufficiently wide variety of genuine use situations for a set of applications may aid the scientific community in the creation of abstraction with promising prospects for real application. Deployments that are dependable are essential for adoption. Implementations must offer tool APIs for efficiency benchmarking, debugging, etc.

8.4. FUTURE TRENDS IN DISTRIBUTED COMPUTING

Our civilization is largely reliant on distributed networks and applications, with networked computers playing a key part in telecommunications, databases, as well as infrastructure management (Carvalho, Cabral, Pereira, and Bernardino, 2021; Mostefaoui, Min, and Müller, 2022; Rashid, Zeebaree, and Shengul, 2019).

8.4.1. Network Architecture

When looking into the future, there are two major issues to consider. The first is to meet the need for ubiquitous connection and frictionless mobility (also known as "Internet as infrastructure") across rival markets and technologies. The second task is to transform today's network infrastructure into "human networks" that are centered on user services. Peer-to-peer, as well as network overlaying methods, might be useful in this case (Affia and Aamer, 2021; Tibrewal, Srivastava, and Tyagi, 2022).

8.4.2. Operating Systems

As is generally true in research and practice, one significant future route is a reimagination of a hot topic from history: virtual machine monitors. Research groups in the United States (such as Denali at the University of Washington) and Europe (such as the XenoServers center at the University of Cambridge) are reactivating VMM methods to advanced frameworks to sustain micro-services, which are elements of Internet-scale programs that demand only a portion of a server's resources. Building small embedded systems for sensor networks (or other very networked systems) as well as the capability for 'ad hoc supercomputing' in the GRID are two more exciting areas for future operating system development (John and Gupta, 2021; Runbang, 2021).

8.4.3. Real-Time Operating Systems

Interconnected systems and operating systems are becoming increasingly more significant shortly. As a result, future real-time information systems would prioritize composability (both large systems and tiny gadgets). Composability is only possible if the required operating system and middleware functionality are available. Proper descriptions of configurable real-time connections are required to make this possible (Baskiyar and Meghanathan, 2005; Malallah et al., 2021; Stankovic and Rajkumar, 2004).

8.4.4. Dependable Systems

The formal formulation of security rules in key "socio-technical" systems to avoid the entrance of flaws and human performance concerns. The state-of-the-art fault-tolerant strategies, particularly in embedded systems, cannot disregard the fact that the majority of failures encountered in actual systems are transitory problems. Fault tolerance in big, large-scale distributed systems, the development of COTS part durability, reflecting technologies to enable introspection and mediation, and modifications to handle security challenges via the idea of intrusion-tolerance are among the other factors (Gray and Cheriton, 1989).

On Fault Removal, the development of statistical testing and robustness testing, and on Fault Forecasting, the establishment of a faithful and tractable model of the system's behavior, the analysis procedures that allow the (possibly very large) model to be processed and the use of fault injection techniques to build dependability benchmarks for comparing competing systems/solutions on an equitable basis (Jadidi, Badihi, and Zhang, 2021; Wu et al., 2021).

8.4.5. The World Wide Web (WWW)

The creation of trustworthy distributed systems known as web Services Architecture is a topic of research that is currently in its early stages. To truly allow the stable construction of Web Services, a variety of research difficulties must be solved. Individual Web Services and their composition must be thoroughly specified to assure the reliability of the resultant systems and to enable the dynamic integration as well as distribution of combined services. Associated dependability methods need also be developed to allow full exploitation of the reliability attributes imposed by particular Web Services, as well as to cope with the Web's unique characteristics. Another challenge is enabling Web Services to be deployed on a variety of devices, from the source of energy smartphones to computers (Figure 8.4) (Colaric and Jonassen, 2021).

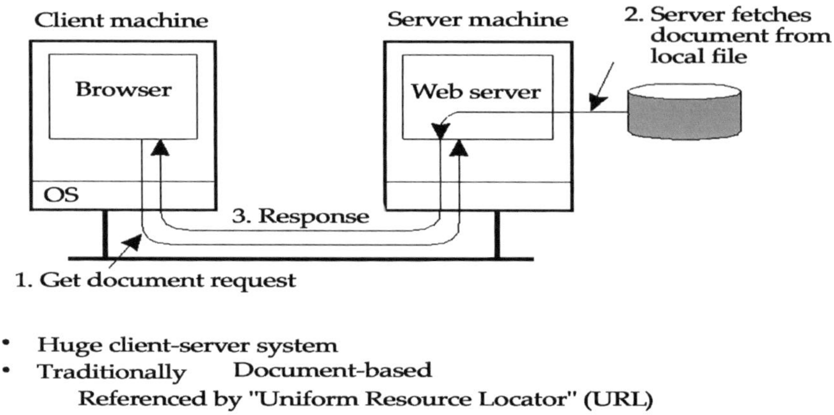

Figure 8.4. World wide web.

Source: https://slideplayer.com/slide/3385366/.

8.4.6. Distributed Event-Based Systems

A fast and reliable infrastructure is required to accommodate the delivery and channel malfunction qualities of a distributed network. Work must proceed on federated systems to enable communication among domain names and event processes within domain names. In addition, password protection (awareness or scope) must be discussed. The large-scale implementation of event processes over multipathing (IP or overlay) to explore the characteristics and issues of high-volume, wide-area event updates is an additional intriguing direction.

8.4.7. Mobile Computing

In the past, a considerable amount of research effort was devoted to improving connectivity services, such as routing protocols, QoS protection, and energy usage. The future improvements will focus on middleware, security, human interfaces, and apps. Specifically, these include (a) virtual cellular connections and ad hoc networks; (b) networking as well as forwarding protocols; (c) development tools for ad hoc networks; (d) power control; (e) security and access control issues; and (f) human interfaces and quality of service.

8.4.8. Network Storage Services

We can anticipate the commoditization of storage as well as the integration of data centers into the network structure. Initial emphasis is placed on logistical connectivity, which entails conceiving of processing and transmission as tightly linked troubles and design solutions correspondingly. Future obstacles include the development of efficient extra storage channels, the improvement of available information during network outages or partitions, the augmentation of data security, and the enhancement of semantic connect connections.

REFERENCES

1. Affia, I., & Aamer, A., (2021). An internet of things-based smart warehouse infrastructure: Design and application. *Journal of Science and Technology Policy Management, 13*(1), 90–109.
2. Asch, M., Moore, T., Badia, R., Beck, M., Beckman, P., Bidot, T., & De Supinski, B., (2018). Big data and extreme-scale computing: Pathways to convergence-toward a shaping strategy for a future software and data ecosystem for scientific inquiry. *The International Journal of High Performance Computing Applications, 32*(4), 435–479.
3. Baskiyar, S., & Meghanathan, N., (2005). A survey of contemporary real-time operating systems. *Informatica (03505596), 29*(2).
4. Carvalho, G., Cabral, B., Pereira, V., & Bernardino, J., (2021). Edge computing: Current trends, research challenges and future directions. *Computing, 103*(5), 993–1023.
5. Cimatti, A., Mover, S., & Tonetta, S., (2011). *Hydi: A Language for Symbolic Hybrid Systems with Discrete Interaction.* Paper presented at the 2011 37th EUROMICRO conference on software engineering and advanced applications.
6. Colaric, S., & Jonassen, D., (2021). Information equals knowledge, searching equals learning, and hyperlinking is good instruction: Myths about learning from the world wide web. In: *The Web in Higher Education: Assessing the Impact and Fulfilling the Potential* (pp. 159–169). CRC Press.
7. Dartnell, L. R., Desorgher, L., Ward, J. M., & Coates, A., (2007). Martian sub-surface ionizing radiation: Biosignatures and geology. *Biogeosciences, 4*(4), 545–558.
8. De Castro, L. N., (2007). Fundamentals of natural computing: An overview. *Physics of Life Reviews, 4*(1), 1–36.
9. Foster, I., Zhao, Y., Raicu, I., & Lu, S., (2008). *Cloud Computing and Grid Computing 360-Degree Compared.* Paper presented at the 2008 grid computing environments workshop.
10. Fox, J. J., & Glass, B. J., (2000). *Impact of Integrated Vehicle Health Management (IVHM) Technologies on Ground Operations for reusable launch vehicles (RLVs) and Spacecraft.* Paper presented at the 2000 IEEE aerospace conference; Proceedings (Cat. No. 00TH8484).
11. Gill, S. H., Razzaq, M. A., Ahmad, M., Almansour, F. M., Haq, I. U., Jhanjhi, N., & Masud, M., (2022). Security and privacy aspects

of cloud computing: A smart campus case study. *Intell. Autom. Soft Comput., 31*, 117–128.

12. Gray, C., & Cheriton, D., (1989). Leases: An efficient fault-tolerant mechanism for distributed file cache consistency. *ACM SIGOPS Operating Systems Review, 23*(5), 202–210.
13. Hwu, W. M., (2014). What is ahead for parallel computing. *Journal of Parallel and Distributed Computing, 74*(7), 2574–2581.
14. Jadidi, S., Badihi, H., & Zhang, Y., (2021). Fault-tolerant cooperative control of large-scale wind farms and wind farm clusters. *Energies, 14*(21), 7436.
15. John, J., & Gupta, S., (2021*). A Survey on Serverless Computing*. arXiv preprint arXiv:2106.11773.
16. Lammie, C., Eshraghian, J. K., Lu, W. D., & Azghadi, M. R., (2021). Memristive stochastic computing for deep learning parameter optimization. *IEEE Transactions on Circuits and Systems II: Express Briefs, 68*(5), 1650–1654.
17. Li, Z., Jiao, B., He, S., & Yu, W., (2022). PHAST: Hierarchical concurrent log-free skip list for persistent memory. *IEEE Transactions on Parallel and Distributed Systems*.
18. Lorenzon, A. F., & Beck, F. A. C. S., (2019). *Parallel Computing Hits the Power Wall: Principles, Challenges, and a Survey of Solutions.* Springer Nature.
19. Malallah, H., Zeebaree, S., Zebari, R. R., Sadeeq, M., Ageed, Z. S., Ibrahim, I. M., & Merceedi, K. J., (2021). A comprehensive study of kernel (issues and concepts) in different operating systems. *Asian Journal of Research in Computer Science, 8*(3), 16–31.
20. Mampage, A., Karunasekera, S., & Buyya, R., (2022). A holistic view on resource management in serverless computing environments: Taxonomy and future directions. *ACM Computing Surveys (CSUR)*.
21. Mansor, M., Harouaka, K., Gonzales, M. S., Macalady, J. L., & Fantle, M. S., (2018). Transport-induced spatial patterns of sulfur isotopes (δ34S) as biosignatures. *Astrobiology, 18*(1), 59–72.
22. Matouš, K., Geers, M. G., Kouznetsova, V. G., & Gillman, A., (2017). A review of predictive nonlinear theories for multiscale modeling of heterogeneous materials. *Journal of Computational Physics, 330*, 192–220.

23. Mitchell, T. M., & Mitchell, T. M., (1997). *Machine Learning* (Vol. 1): McGraw-Hill New York.
24. Mostefaoui, A., Min, G., & Müller, P., (2022). *Connected Vehicles Meet Big Data Technologies: Recent Advances and Future Trends.* In: Elsevier.
25. Murad, S. A., Muzahid, A. J. M., Azmi, Z. R. M., Hoque, M. I., & Kowsher, M., (2022). A review on job scheduling technique in cloud computing and priority rule based intelligent framework. *Journal of King Saud University-Computer and Information Sciences.*
26. Navau, C., & Sort, J., (2021). Exploiting random phenomena in magnetic materials for data security, logics, and neuromorphic computing: Challenges and prospects. *APL Materials, 9*(7), 070903.
27. Patty, C. L., Ten, K. I. L., Buma, W. J., Van, S. R. J., Steinbach, G., Ariese, F., & Snik, F., (2019). Circular spectropolarimetric sensing of vegetation in the field: Possibilities for the remote detection of extraterrestrial life. *Astrobiology, 19*(10), 1221–1229.
28. Pigatto, D. F., Gonçalves, L., Pinto, A. S. R., Roberto, G. F., Rodrigues, F. J. F., & Branco, K. R. L. J. C., (2014). *HAMSTER-Healthy, Mobility and Security-Based Data Communication Architecture for Unmanned Aircraft Systems.* Paper presented at the 2014 International Conference on Unmanned Aircraft Systems (ICUAS).
29. Rashid, Z. N., Zeebaree, S. R., & Shengul, A., (2019). *Design and Analysis of Proposed Remote Controlling Distributed Parallel Computing System Over the Cloud.* Paper presented at the 2019 international conference on advanced science and engineering (ICOASE).
30. Ravi, S., Raghunathan, A., Kocher, P., & Hattangady, S., (2004). Security in embedded systems: Design challenges. *ACM Transactions on Embedded Computing Systems (TECS), 3*(3), 461–491.
31. Runbang, L., (2021). *Design and Implementation of Neural Computing Platform Application System Based on Super Computing.* Paper presented at the 2021 international conference on Intelligent transportation, big data & smart city (ICITBS).
32. Sadiku, M. N., Musa, S. M., & Momoh, O. D., (2014). Cloud computing: Opportunities and challenges. *IEEE Potentials, 33*(1), 34–36.

33. Saini, K., Kalra, S., & Sood, S. K., (2022). An integrated framework for smart earthquake prediction: IoT, fog, and cloud computing. *Journal of Grid Computing, 20*(2), 1–20.
34. Sarker, I. H., (2021). Machine learning: Algorithms, real-world applications and research directions. *SN Computer Science, 2*(3), 1–21.
35. Schuld, M., Sinayskiy, I., & Petruccione, F., (2014). The quest for a quantum neural network. *Quantum Information Processing, 13*(11), 2567–2586.
36. Schultz, M., Rosenow, J., & Olive, X., (2022). Data-driven airport management enabled by operational milestones derived from ADS-B messages. *Journal of Air Transport Management, 99*, 102164.
37. Skoglund, M., (2022). *Towards an Assessment of Safety and Security Interplay in Automated Driving Systems.* Mälardalen University.
38. Stankovic, J. A., & Rajkumar, R., (2004). Real-time operating systems. *Real-Time Systems, 28*(2, 3), 237–253.
39. Sunderam, V. S., (1990). PVM: A framework for parallel distributed computing. *Concurrency: Practice and Experience, 2*(4), 315–339.
40. Sutar, S., & Mekala, P., (2022). An extensive review on IoT security challenges and LWC implementation on tiny hardware for node level security evaluation. *International Journal of Next-Generation Computing, 13*(1).
41. Tan, L., Kim, J., Sokolsky, O., & Lee, I., (2004). *Model-Based Testing and Monitoring for Hybrid Embedded Systems.* Paper presented at the proceedings of the 2004 IEEE international conference on information reuse and integration. IRI.
42. Tchernykh, A., Schwiegelsohn, U., Talbi, E. G., & Babenko, M., (2019). Towards understanding uncertainty in cloud computing with risks of confidentiality, integrity, and availability. *Journal of Computational Science, 36*, 100581.
43. Tibrewal, I., Srivastava, M., & Tyagi, A. K., (2022). Blockchain technology for securing cyber-infrastructure and internet of things networks. *Intelligent Interactive Multimedia Systems for e-Healthcare Applications*, 337–350.
44. Wilkins-Diehr, N., Gannon, D., Klimeck, G., Oster, S., & Pamidighantam, S., (2008). TeraGrid science gateways and their impact on science. *Computer, 41*(11), 32–41.

45. Wu, Q., Xu, J., Zeng, Y., Ng, D. W. K., Al-Dhahir, N., Schober, R., & Swindlehurst, A. L., (2021). A comprehensive overview on 5G-and-beyond networks with UAVs: From communications to sensing and intelligence. *IEEE Journal on Selected Areas in Communications*.
46. Wu, Y., Liu, D., Chen, X., Ren, J., Liu, R., Tan, Y., & Zhang, Z., (2021). MobileRE: A replicas prioritized hybrid fault tolerance strategy for mobile distributed system. *Journal of Systems Architecture, 118*, 102217.
47. Xu, Z., & Wu, J., (2022). Computing flow pipe of embedded hybrid systems using deep group preserving scheme. *Cluster Computing, 25*(2), 1207–1220.
48. Yousif, A., Alqhtani, S. M., Bashir, M. B., Ali, A., Hamza, R., Hassan, A., & Tawfeeg, T. M., (2022). Greedy firefly algorithm for optimizing job scheduling in IoT grid computing. *Sensors, 22*(3), 850.
49. Zarei, A., Safari, S., Ahmadi, M., & Mardukhi, F., (2022). *Past, Present and Future of Hadoop: A Survey*. arXiv preprint arXiv:2202.13293.
50. Zeng, H., (2022). Influences of mobile edge computing-based service preloading on the early-warning of financial risks. *The Journal of Supercomputing, 78*(9), 11621–11639.
51. Zomaya, A. Y., (1996). *Parallel and Distributed Computing Handbook(pp. 1-5).*

INDEX

O

P

Q

R

S

T

U

V

W

X